THE LAST CHANCE
CANAL COMPANY

Charles
Redd
Monographs
in Western
History No.16

THE LAST CHANCE CANAL COMPANY

Max R. McCarthy

Brigham Young University
Charles Redd Center for Western Studies

The Charles Redd Monographs in Western History are made possible by a grant from Charles Redd. This grant served as the basis for the establishment of the Charles Redd Center for Western Studies at Brigham Young University.

Library of Congress Cataloging-in-Publication Data

McCarthy, Max R., 1919–
  The Last Chance Canal Company.

  Bibliography: p. 129
  1. Last Chance Canal Company (Idaho)—History.
2. Irrigation canals and flumes—Idaho—History.
3. Irrigation—Government policy—West (U.S.)—History.
I. Title.
HD1739.I2M32 1986   338.7′6317′09796   86-23295
ISBN 0-941214-53-2

Distributed by Signature Books, Salt Lake City, Utah

Table of
Contents

Preface and
Acknowledgments

The chronicles of the West are replete with dramatic "last chance" episodes. But perhaps in few instances is the term more significant, and more enduring, than in the Gem Valley of southeastern Idaho. For it was here, nearly nine decades ago, that a group of mostly Mormon settlers staked their dreams for the future on an irrigation enterprise called the Last Chance Canal Company.

How they successfully met the formidable challenges of devising a system whereby they could irrigate their semiarid acres with water from the Bear River is a story worthy of the finest pioneer traditions. And the ironic twist that characterizes the ultimate triumph is a tribute to the resourcefulness, perseverance, and sense of destiny that were marks of these early settlers and those who followed them.

The name Last Chance Canal Company was first associated with the commonly regarded "last chance" attempt in 1897 to irrigate from Bear River. In 1981 the "Last Chance" portion of the name was added to a subsidiary activity, the Last Chance Hydro Electric Company.

Irrigation was essential to successful agriculture in Gem Valley. This premise dominates both the history and the folklore of that area, making the eventual solution to the irrigation problem both a major achievement and a source of great local pride.

I was raised in Gem Valley. My general awareness of the vital role of the Last Chance Canal Company was reinforced by several special considerations. My grandfather, L. B. McCarthy, granted his homestead in 1904, was an early settler in what later became Turner and was an early stockholder in the Last Chance Canal Company. My father, John Roy McCarthy, continued to farm the homestead and other Gem Valley properties and was a longtime officer of the Turner Canal Company, an element of the Last Chance system. As for me, I worked during two summers while in college as the "ditch rider," or watermaster, patrolling the canals on horseback along two of the Last Chance system canals.

Perhaps because of the family affiliation and my personal involvement with the Last Chance irrigation system, I developed early a firmly held conviction that the hydroelectric facilities of the Utah Power and Light Company (UP&L) in Gem Valley operated to the disadvantage of the local farmers. This perception of an adversary relationship is largely responsible for my decision to investigate matters more thoroughly and test the validity of my perception. In this pursuit I found that an adequate written history of the Last Chance system did not exist. To correct this deficiency is the purpose of the narrative that follows.

The availability of source material is of vital importance to support any effort at historical narrative. I have relied on secondary sources in preparing the chapters that are mostly background in nature. However, I have used primary sources not generally available in addressing the Last Chance irrigation system itself and the court case by which the waters used in its operation were adjudicated. Primary sources thus support the subjects of principal significance. Although exhaustively researched, these sources present one problem; they are not, in all instances, chronologically complete. Individuals associated with canal construction and early canal company decisions were not motivated solely by the future historical implications of their work, and the records are occasionally incomplete or have been lost or destroyed.

Many individuals and offices have aided research on this project over a period of several years. Among those to whom I extend my gratitude and appreciation are: Cecil Alldaffer, then extension agent, Caribou County, Soda Springs, Idaho; Randall C. Budge, attorney for the Last Chance Canal Company, Pocatello, Idaho; Joyce Fowler, librarian, Public Library, Grace, Idaho; Jeanie M. Loera, deputy clerk, U.S. District Court, Boise, Idaho; Ralph J. Mellin, Department of Water Resources, State of Idaho, Boise; Laura M. Pershing, law librarian, State of Idaho, Boise; A. J. Simmonds, Special Collections Librarian, Milton R. Merrill Library, Utah State University, Logan, Utah; and Russell D. Stoker, Bear River watermaster, Soda Springs.

I also acknowledge invaluable assistance from Keith R. McCarthy, my brother and also a former Last Chance system "ditch rider," then an employee of the Department of Interior, Washington, D.C., who gave me initial encouragement, arranged many valuable contacts for me within agencies of the federal government and the state of Idaho, and who also reviewed this manuscript.

Without the friendly, enthusiastic cooperation of Orrin Harris, then secretary, Last Chance Canal Company, Grace, Idaho, in all probability this story could not have been written. Mr. Harris opened his office and his files for my research, cheerfully corrected any erroneous interpretations of data on my part, and escorted me on several occasions to tour Last Chance facilities.

Finally, I accord special acknowledgment and appreciation to my wife, Oretta, who invariably encouraged my efforts, tolerated with patience my many trips to research the project, and who typed this effort through its several versions.

Introduction

This narrative recounts significant events occurring primarily within southeastern Idaho as they affected practices and policies regarding a scarce natural resource—water—both in that state's early, developmental stages and later. Broadly, the area of concern is much of the Bear River watershed. More specifically, I focus on development within one area— Gem Valley—of that watershed.

The location and importance of the watershed have been known since the discovery of its major river in the summer of 1812 by a small party of the Wilson Price Hunt land expedition of Astorians returning eastward from Oregon.[1] Following this discovery, fur-trapping activities dominated the area for the next quarter century. In 1818–19 Donald McKenzie named the two principal geographic features of the watershed Bear River and Bear Lake after recording the presence there of a great many black bears.[2] Another trapper and early explorer, Jim Bridger, was in Bear River Valley in the early 1820s as a member of an Ashley trapping party. During one foray to the south he passed down a portion of the river to its mouth and discovered the Great Salt Lake as the drainage basin of the Bear River.[3] Another great fur trapper, Jedediah Smith, was in charge of the trappers' 1827 rendezvous held at Bear Lake.[4]

By the mid-1830s, the dominance of the fur trappers was beginning to wane. Because the valley of the Bear River offered easy passage through some of the mountain ranges of the region, it inevitably became an important link in cross-continent travel routes. One heavily used route within the watershed area, a segment of the famous Oregon Trail, ran from Fort Bridger northwest down the Bear River Valley to Soda Springs, and thence generally west-northwest to Fort Hall on the Snake River.

The growing number of travelers trekking westward on this trail were a varied lot. While the number of trappers dwindled, missionaries and

1

emigrants en route to Oregon became more common. Samuel Parker and Marcus Whitman, Presbyterian missionaries accompanied by their wives, moved across the trail in 1836.[5] By 1843 large numbers of emigrants were using this trail, and the "westward ho" movement became the "Great Migration."

The accumulated knowledge about the Bear River and Bear Lake geographic area was available from several sources. Washington Irving's book, *The Adventures of Captain Bonneville,* published in 1837, included a description of Bonneville's presence at the lake and along the river in 1833.[6] There was also word-of-mouth advertising "back East" by missionaries who had earlier moved across the Oregon Trail.[7] A final major effort in publicizing the West before arrival of large numbers of settlers came from John C. Frémont and his "great reconnaissance" of 1843–44. This exploratory venture brought him to the Bear River in late August 1843, and later to Salt Lake Valley. His official reports of his explorations gave the first comprehensive appraisal of the Great Basin,[8] of which the Bear River watershed was a part.

The Mormons used these Frémont reports in particular in planning their westward exodus. From their studies of the Frémont reports and other materials, church leaders concluded that their destination beyond the Rocky Mountains was to be either the vicinity of the Great Salt Lake or Bear River Valley.[9]

The Mormons selected the Salt Lake Valley as the location for their initial settlement in 1847, and locales within Idaho in Bear Lake Valley and along the Bear River were settled fifteen to twenty years later as a result of northward expansion from the Salt Lake center. By the early 1860s settlements had advanced to Franklin, Bear Lake Valley, and Morristown (present-day Soda Springs). Additional locations within the Bear River watershed were occupied over the next thirty years, and settlement of the somewhat inhospitable Gem Valley became a *fait accompli*.

The success of settlements in Gem Valley largely depended on successful development of the irrigation ventures that became known in local idiom as the "Last Chance," and more formally as the Last Chance Canal Company. Also, within twenty years the great potential of Bear River for hydroelectric development was recognized. The generally simultaneous irrigation and hydroelectric development of the Bear River watershed constitute much of the more recent history. The irrigation story, in major part, is a tale of pioneers in the best tradition of the

American West. Present are elements of hardship, stubborn resolve, visionary leadership, fierce competition for the precious water resource, and eventual triumph. However, the ending to the story is unique, in some ways ironic. The two principal protagonists—the irrigation interests and the power interests—both finally succeed, and their separate successes bring mutual respect and friendship. Thus, after more than eight decades, the two groups acting out the drama have come center stage to take a common bow, with the final act having been highlighted by an ironic reversal of roles with the Last Chance demonstrating its ability to produce electric power as well as provide irrigation.

Background material about the environment and circumstances in which the Last Chance system developed and functioned provides a fuller understanding of the Last Chance story. This background information includes chapters dealing with the Bear River watershed, irrigation knowledge extant at the time Gem Valley was settled, applicable Idaho water law, earlier unsuccessful irrigation efforts, and competing hydroelectric developments. In this latter instance, there is no pretense of presenting a comprehensive history of UP&L or its predecessors, the corporations responsible for such developments. The intent is solely to establish the environment and circumstances affecting the Last Chance system.

And it is against this background that the role of the Last Chance irrigation system unfolds. The story begins with chapters on the Last Chance irrigation project and on the Bear River Water Case. A following chapter covers the more recent history and deals with challenges and solutions. The volume concludes with an assessment of the overall significance of the Last Chance episode.

To summarize, within five years of the first appropriation of Bear River water by the Last Chance system, irrigation of Gem Valley lands was possible. The system expanded over the years under capable, innovative leadership, and became the basis for an exceptionally fine agricultural enterprise. But throughout a crucial concern persisted: consistent availability of enough water. The Dietrich decree in 1920 adjudicated water to the Last Chance system basically equal to the capacity of their canals. However, in years of drought when the natural river flow was low, expensive supplemental water was required. The source was the Bear Lake Reservoir, which was controlled by UP&L. Rent of this supplemental water was an economic burden to the farmers and one under which they had chafed for many years.

The eventual solution came in an unusual and perhaps surprising form. In an action unrelated to the Last Chance's traditional preoccupation with irrigation, the subsidiary Last Chance Hydro Electric Company planned and built its own hydroelectric plant between 1980 and 1983. The Last Chance Canal Company traded this plant to the power company in January 1984 for a permanent supply of supplemental irrigation water from the Bear Lake Reservoir, at no cost, in addition to other considerations. Thus, the final irony: the Last Chance organization attained its long-sought guarantee of sufficient irrigation water by developing the ability to generate electricity, the forte of its historical competitors.

The story of the Last Chance organization in Gem Valley is a case study in the evolvement of a broader water resource policy in the West. The influence of the Last Chance microcosm on the Bear River extends beyond its direct geographical area. Developments there are illustrations of Idaho water policy as it affected both irrigation and hydroelectric interests. The "Bear River Water Case" of 1917–20 and the famous "Dietrich Decree" by which it was settled are among the earliest legal precedents in adjudicating water rights. These events affected later adjudication actions in Idaho courts and the eventual formulation of the means to coordinate competing water interests between states.

Indeed, what started out to be an isolated pioneer effort to make possible the agricultural development of southeastern Idaho lives on as an important adjunct to western water policy, as well as a local success story of commendable and enduring proportions.

CHAPTER ONE

The Bear River
Watershed

The Bear River watershed, occupying an area of limited precipitation and covering about six thousand square miles, provides drainage for an estimated 1.54 million acre-feet of water annually.[1] A major problem with this significant quantity of water is the seasonal variation in its availability.

Although there are numerous springs, creeks, and tributary rivers, the two principal hydrological features of this drainage area are Bear Lake and Bear River. Bear Lake, lying astride the Utah-Idaho boundary in the center of a large, marshy tract, is about twenty-one miles in length and six to eight miles in width. This lake originated with the Bannock overthrust and the dropping of the valley floor. The resulting depression later filled with water. The lake is now about two hundred feet deep at its deepest points with an average depth of thirty to forty feet.[2] The elevation of the water surface of Bear Lake varies seasonally and from year to year. In 1919 the elevation was described as between 5,919 and 5,923 feet, varying according to the season.[3] In late 1983 the elevation was 5,922.2 feet. The "all-time high," first recorded in 1922, was 5,923.65 feet.[4] Dike construction now limits the maximum lake elevation to the 1922 "high" figure. When water fills the lake to that level, excess water is manually released to return to Bear River—as was necessary in 1986.

The relationship of Bear Lake to Bear River has been an important consideration in litigation over water rights. The "river does not flow in or out of it [Bear Lake] naturally but passes by several miles to the north east."[5] However, "in ages past, Bear River fed directly into Bear Lake before continuing its flow into Great Salt Lake."[6] This river passage through Bear Lake was interrupted on several occasions. A temporary blocking of the river below the lake, perhaps by glacial activity out of Georgetown Canyon, is advanced as an explanation of the terracing above the present lake-surface level on some of the hills rimming the lake.[7] Probably at a later time, deposition of silt may have created an embankment that "diverted the river along the north shore of the lake."[8]

Even after the relocation of the river channel to the northeast of Bear Lake, there remained an outlet from the lake to the river. The old outlet traced a "torturous course" northward to its junction with the river. As the outlet passed northward from Bear Lake, it entered a water area, partially spring fed, called Mud Lake or North Lake, and then passed through a swamp north of Mud Lake finally to join the Bear River. Fall along this outlet of 4 to 7.6 feet is recorded.[9]

Water flow through this outlet is a matter of controversy. To an early resident of the area, it seemed that "in the spring the river would overflow, and the creeks and floods from the mountains would come down and flood the valley, and the water would run into the lake . . . until the lake was full, and the outlet had drained the valley, or sloughs, to such an extent that the water would then start from the lake. . . . The water would run steadily and slowly during the summer season until spring came again. There was a steady flow out from the storage of the [lake]."[10]

A contrasting view of the possibility of water flow through the natural connection between Bear Lake and Bear River asserts that "only during flood stage did it [Bear River] overflow . . . [the] silt embankment and find its way back into Bear Lake via the sloughs which were named Mud Lake."[11] This view maintained that the water flow was both sporadic and one-way from the flooded river to the lake. Obstruction to the channel (the so-called natural outlet) made it useless as a "conduit for water." The "non-contributory character of the lake to the river" during the summer season was proclaimed.[12]

Thus far, the only characteristic of the Bear River that has been established is that the river does not flow through or clearly join Bear Lake. Other aspects of the river are also of interest. Bear River has its source in Amethyst and McPheters lakes on the north slope of Hayden Peak (12,485 feet elevation), near the western limit of Utah's High Uintas. From this origin the stream passes northward into Wyoming just inside the western boundary. North of Evanston the river snakes back into Utah, back again into Wyoming, and then into Idaho near the town of Border, Wyoming. From this location, the river continues in a northwesterly direction past the north end of Bear Lake, past Soda Springs, and on to a point near Alexander, Idaho.[13]

At about this point the river channel has changed. Before approxi-

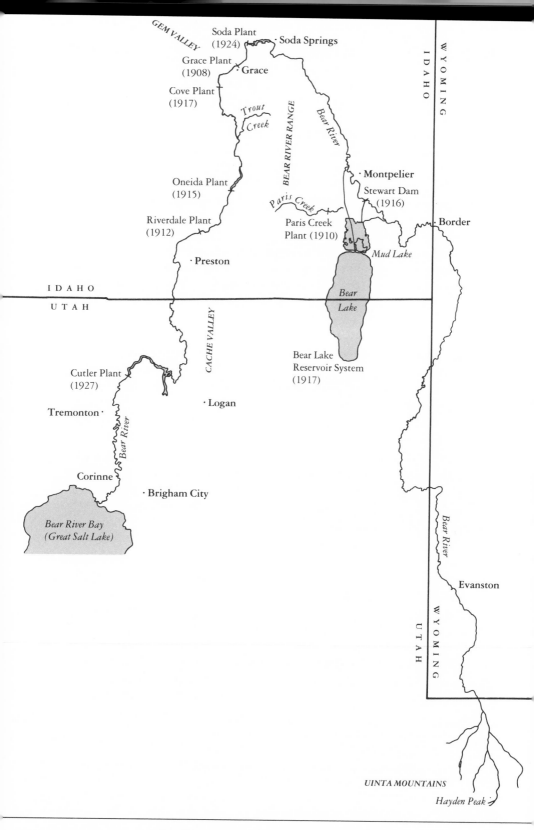

Bear River Watershed and Hydroelectric Plants

mately thirty-four thousand years ago, the Bear River continued its northwesterly course to empty into the Portneuf River, its waters eventually reaching the Pacific Ocean via the Snake and Columbia rivers. Then the Bear River was blocked by a lava flow from the northeast. Ancient Lake Thatcher was formed in lower terrain south of present-day Grace, Idaho, its water finally spilling over the divide into Cache Valley to become a part of the Lake Bonneville or Great Basin drainage.[14]

As the Bear River follows its present course southward through Idaho and Utah until it empties into Great Salt Lake, the river direction has, from the direction of flow at its source, virtually doubled back on itself. Washington Irving described this great bend of the Bear River as the "figure of a horseshoe,"[15] or to illustrate this point differently, Bear River follows a course five hundred miles long but its mouth is only ninety miles from its source.[16]

The course of the river presents two characteristics important to altering or controlling its flow. First, a "canyon section of the river" runs from Gentile (Gem) Valley to the upper end of Cache Valley to the south.[17] Second, from survey data available at the time of river development, of the total river fall of 1,720 feet from the Bear Lake to Great Salt Lake, approximately 84 percent (1,439 feet) occurs within the constrictions of this canyon section. Of particular appeal to future hydroelectric developments was the "canyon section fall," which was identified with specific river locations at Soda (79 feet), Alexander (89 feet), Grace (525 feet), Cove (94 feet), Oneida (145 feet), Mink Creek (115 feet), and Cutler (128 feet).[18]

Whereas the characteristics of Bear River through the canyon section were especially suitable to power development, the existence of the rocky canyon gorge through much of Gem Valley enormously complicated any effort to get irrigation water out onto the surrounding fertile land. Thus, settlement of the upper Gem Valley areas was delayed as compared to the Bear Lake Valley and the Cache and lower Gem valleys.

Another characteristic of the Bear River before settlement—much different than the present—is size. Before the river had been manipulated for power or irrigation purposes, it was a stream of significant flow. An early photograph of the river near Grace clearly shows a depth of water within a few feet of the canyon rim. The photo caption further illustrates the river's size. "Those who lived here then will remember the roar of the river could be heard anywhere in the valley, night or day."[19]

It is clear from settlement experience in the American West that water presence and the ease with which it could be applied to the soil were important factors in both locations and dates of settlement. This was certainly true throughout the course of the Bear River. However, if considered in combination, Bear Lake and Bear River together provided a singular promise for the future of this portion of the Great Basin. To the imaginative and resourceful, a "vision" existed of the "feasibility of great Bear River developments reinforced by the reserves of the Bear Lake Reservoir." The lake, "claimed to be the largest natural reservoir suitable for irrigation purposes in the world," was visualized as a place "in which could be stored the flood waters of Bear River, which, when released, could be used for irrigation and for the generation of power."[20]

CHAPTER TWO

Prospective Settlers'
Introduction to Irrigation

There were several means by which the area's prospective settlers after mid-century could have been informed about irrigation techniques, engineering, prospects, and legalities. Two such means were experiences in irrigation from Mormon settlements centered in Salt Lake Valley and the publicity incident to the "national irrigation movement" of the late nineteenth century.

Probably of greatest importance to the eventual settlers of Gem Valley was a knowledge of Mormon irrigation experiences—a knowledge easily accessible since those settlers originated from Utah. Mormon leaders had been actively involved in irrigation matters even before their arrival in Utah.[1] And once the pioneer party reached the selected location near Great Salt Lake in 1847, they immediately started irrigating. One day after arrival, what is now called City Creek was diverted onto the land to soften it for plowing, and the next day, July 24, potatoes were planted and irrigated. As the Mormon settlement, Great Salt Lake City, was platted, provisions were made for a major canal following present-day North Temple Street west to the Jordan River. Laterals from this canal conveyed water separately to each of the ten-acre blocks into which the settlement was divided. These extensive plans for irrigation were modified as required by new circumstances and applied in other areas as the colonization expanded from the initial settlement.[2] Indeed, water for irrigation was a primary determinant in the pattern of settlement.[3]

Since irrigation was at first practiced mostly on bottom lands, Mormon irrigation techniques and facilities were simple. Diversion dams were frequently constructed only of rocks and brush. Ditches were short, narrow, and shallow. Robert G. Dunbar provides an interesting description of their construction.

> Lacking sophisticated surveying instruments, the ditch builders
> sighted over a pan of water or determined the proper gradient by the

11

use of a triangular frame with a plumb line suspended from its apex. Once the route of the ditch had been determined, the builders dug the ditch by plows, go-devils, shovels, and spades. The go-devil was an A-shaped implement made of logs, which, when drawn by oxen, pushed the plowed dirt to the sides and hollowed out the ditch.[4]

Several features characterized this early Mormon irrigation. First, water was regarded as belonging to the community; private ownership of this natural resource was not visualized. Second, construction of the irrigation facilities in this period was a small-scale operation. Each settler contributed labor in proportion to his holdings. And third, the church exerted an influence—changing through the years from direct to a more subtle form—over the irrigation planning and operation. Initially, watermasters were appointed by the bishops and water controversies were handled in bishops' courts. However, church influence lessened as municipal and other governmental bodies were established.[5]

Although early Mormon irrigation techniques have been criticized as expedient solutions utilizing the "crudest devices, and with no thought for any grand and enduring scheme of engineering,"[6] an awareness of the Mormon experience was critically important to prospective settlers. John Wesley Powell commended specifically two characteristics as contributing to the success of irrigation in Utah Territory—the ecclesiastical organization and the application of cooperative labor.[7] William E. Smythe noted further that cooperative labor was virtually the only means of obtaining water rights (for example, canal stock) in an irrigation activity.[8]

Other events possessing the potential to stimulate further interest in irrigation began after about a quarter century of Mormon irrigation experience. In 1877 Congress enacted the Desert Land Act, a measure specifically designed to encourage irrigation. Under this measure, U.S. citizens in designated western states could purchase one section (640 acres) of "desert lands" of the public domain for $1.25 per acre, provided irrigation of one-eighth of the claim was accomplished within three years. Unfortunately, most of the land remaining open for settlement at that time required complex irrigation systems that would take more than three years to complete. Further, considerable fraud in the form of "absentee entrymen" became associated with this act. Nonetheless, an assessment was that "there is no question but that the Desert Land Act encouraged irrigation."[9]

Attention was further directed toward irrigation in 1879 when a document, *Report on the Lands of the Arid Region of the United States: U.S. Geographical and Geological Survey of the Rocky Mountain Region,* was published. This report, compiled under the supervision of John Wesley Powell, was intended to classify all public lands of the region as to aridity, to map and measure streams, to locate suitable reservoir sites, and to evaluate area climate.[10] The Bear River was among the large streams specifically analyzed by Powell. He traced its course and observed that proper distribution of the water among potential competing users would result in "reclamation of the best selection of land." He also recorded his estimate that 15 percent of the total area of the "Bear River District" was susceptible of cultivation, but that only 2.5 percent was actually under cultivation. Irrigation was, in Powell's view, the prerequisite to reclaiming a larger percentage of the "cultivable" land.[11] On a scale broader than merely the Bear River watershed, Powell visualized "the deserts gone, the waste areas reclaimed."[12]

Powell's report presaged the national irrigation movement. Congressional interest had been attracted. First, Congress authorized the study of specific reservoir sites in 1888. Two years later a congressional committee toured the arid regions, conducting public hearings at numerous points.[13] Coincident with this exploratory and tentative interest in irrigation, in 1890 a severe drought occurred on the Great Plains. This "terrible misfortune . . . put active life into the irrigation movement."[14] Practical considerations also stimulated interest in irrigation-related activities. Irrigators began to see the benefits to result from reservoir development. For example, storage of unneeded water would reduce waste, more land could be placed under cultivation, and the irrigation season could be extended to accommodate late-maturing crops.[15]

The first demonstration of this intensified interest in irrigation involved private rather than governmental response. In September 1891 the first National Irrigation Congress was held at Salt Lake City. The objectives of this and later congresses were to obtain cession of arid but irrigable lands from the federal government to the states, to reform state water law, and to encourage irrigation investment possibilities.[16] These objectives, stressing state involvement in water resource development as they did, were important—if for no other reason than costs of construction of irrigation works and reservoirs were high (and rising) and the burden would have to be shared at all levels of government.

As an "eloquent voice for national reclamation," the journal *Irrigation*

Age was also established at this time. William E. Smythe, the first—and very capable—editor, espoused irrigation not solely as an adjunct to agriculture, but as a "philosophy, a religion."[17] Irrigation was advanced as an "instrument for transforming society in the western third of the continent."[18] A second congress, held at Los Angeles in 1893, proclaimed the irrigation question to be national in essence. The next year the irrigation congress convened in Denver and called for increased appropriations for investigation of water supplies and the creation of a national commission to develop plans for reclamation of arid lands. The next five years saw congresses held at Albuquerque, Phoenix, Lincoln, Cheyenne, and Missoula.[19] The 1896 congress adopted a resolution favoring federal construction of storage reservoirs.[20]

The educational efforts of the irrigation congresses, held at diverse locations throughout the arid regions, were felt within the federal government. Irrigation had become an aspect of the national conservation concern—a concern of broad appeal to many in the Progressive Era. On August 17, 1894, legislation that came to be known as the Carey Act from its sponsor, Senator Joseph M. Carey of Wyoming, was enacted. Under this act the federal government was to provide one million acres of arid desert lands to each of eleven western states and territories, if certain prescribed measures concerning irrigation were taken in each case. In turn, the government visualized development corporations using private capital to construct the required irrigation facilities under careful supervision from the states or territories. Expenses charged to each prospective settler included fifty cents per acre for the authorized 160-acre tract and a proportional allocation of the cost of the irrigation project, amounts that varied from project to project, but were estimated at ten to twenty-five dollars per acre. To curb speculation, purchase of the water rights was prerequisite to the land sale. With each settler's purchase of water rights went a share in a mutual irrigation company, to which project management and operation was transferred when the construction was completed and all rights to the water sold. Land sale could be completed as soon as one-eighth of the tract had been irrigated.

There were limited benefits to irrigation from the Carey Act, however. Development corporations frequently miscalculated costs of the irrigation projects, water supply was often overestimated, and there seemed at times to be a dearth of settlers to purchase the water rights.[21] The results from the Carey Act were considered disappointing.[22]

Two events significant to the cause of irrigation occurred in 1897. One involved the establishment of a formal, continuous organization to strengthen the already existing, but informal, National Irrigation Congress structure. To this end the National Irrigation Association, an avowed pressure group, was established at Wichita, Kansas, under the vigorous leadership of George H. Maxwell.[23] During this same year a report, to prove almost as influential as Powell's report twenty years earlier, was completed by Captain Hiram M. Chittenden. The report, *Reservoirs of the Arid Region,* was a clear call for direct federal participation in irrigation for the arid regions. Chittenden recommended that the federal government acquire full title and jurisdiction over suitable reservoir sites and full rights to the water available to fill the reservoirs. In this first report devoted specifically to reservoirs, Chittenden identified three sites in Wyoming and two in Colorado. His intent was that the federal government build, own, and operate the reservoirs. Water stored therein was to be free for public use under local regulations. Flood control was an added advantage in the areas to be served by the reservoirs.[24]

The year 1900 also offered two significant occurrences concerning irrigation. One was the appearance of William E. Smythe's book, *The Conquest of Arid America.* This book, which has been called a "classic of the Progressive Era reclamation movement,"[25] became very popular and was of major assistance in achieving recognition of irrigation as a national rather than merely regional concern. The other occurrence was that the platforms of both of the major political parties in the 1900 national elections endorsed the concept of reclamation of arid lands.[26]

The irrigation movement, which had been the subject of much private and governmental attention for twenty-five years, culminated with the enactment of the Reclamation Act of 1902. This act, popularly called the Newlands Act after its sponsor, Representative (later Senator) Francis G. Newlands of Nevada, provided for the direct government construction of irrigation projects to be paid for from the total revenues from the sale of public lands in the West. The newly created Reclamation Service within the Department of Interior was responsible for administering government involvement in irrigation. Frederick H. Newell was the first director.

The Reclamation Service was charged first with the examination and survey of the streams and surrounding land that was susceptible to irrigation, then the construction and maintenance of irrigation works for the storage, diversion, and development of water for reclamation of arid and semiarid lands. The area of responsibility included thirteen states and three territories. Title to reservoirs and works developed under the 1902 act remained with the government, but after the government had been reimbursed for its involvement, maintenance and operation could be passed to the landowners. Reimbursement to the government came in payment for land at $1.25 per acre and payment of an amount fixed by the Secretary of Interior for the irrigation works. This latter amount was due in ten annual installments.

The first five projects approved under the provisions of the Reclamation Act of 1902 were on the Salt River in Arizona, the Gunnison River in Colorado, the Milk River in Montana, the Truckee River in Nevada, and the Sweetwater River in Wyoming. Funds for these projects, set aside from public land sale revenues, were subject to a continuing appropriation and were placed in a special fund for reclamation. The fund grew rapidly to sixteen million dollars the first year after enactment and to thirty million in 1905.[27] By 1907, twenty-four projects, most involving reservoirs, diversion dams, and distributing canals, had been approved in fifteen states.[28]

The act visualized a concept of multipurpose projects. In addition to irrigation, concerns were expressed for navigation, flood control, and possible generation of hydroelectric power.[29] In particular, an awareness of the fundamental importance of sites for dams, including hydroelectric sites, was reflected.[30]

Through the Reclamation Service, whose precise role was still largely undefined, the federal government was clearly to be an active participant in future irrigation activity. But primacy of the irrigation role was not assured. Governmental irrigation interests were often not totally in sympathy with the largely privately financed hydroelectric, timber, mineral, and livestock interests in the West. Further, a keen competition for control of water existed between those interested in its use for irrigation and those whose emphasis was hydroelectric power. Within the Bear River watershed, resolution of this competition would require much expense and attention.

In summary, government reports, federal laws, and private publications and congresses brought national awareness to all aspects of irrigation by the end of the nineteenth century. Many Gem Valley settlers would have had firsthand experience applying Mormon irrigation procedures as well. Irrigation was a major concern in the livelihood of these settlers, and with national and local attention focused on irrigation, state involvement would soon follow.

CHAPTER THREE

Idaho Water Policy

Some aspects of Idaho governmental involvement in irrigation are obvious upon reviewing the applicable territorial and state law. Since the period of concern to this study is limited roughly to the period 1885 through 1920, the law documents particularly meriting attention are the Idaho statutes for 1887,[1] 1895,[2] 1899,[3] and 1903,[4] and the Idaho State constitution of 1890.[5] There are, of course, some fluctuations among these documents in minor detail. However, a careful comparison of the statutes for repetitive, significant provisions discloses a consistent pattern of principles and procedures that may properly be termed "Idaho water policy."

A summary of portions of this water policy is prerequisite to further discussions of the legal constraints on irrigation and hydroelectric developments and of the major court battles among the competitors for Bear River water. To provide this summary is the intent of this chapter.

Basic to any consideration of Idaho water policy is the concept that the "waters of the State are held to belong to the State."[6] This "title to the public waters of the State is vested in the State for the use and benefit of all citizens under such rules and regulations as may be prescribed from time to time by the legislature."[7]

The right of citizens to the use of these public waters is acquired by appropriation, such appropriation being completed by actually diverting water from its natural watercourse and applying it to "a beneficial use."[8] Eligibility to appropriate water was extended to "any person, association or corporation" desiring to do so.[9] "Beneficial use" is defined as use for domestic, agricultural, and manufacturing purposes, in that order of priority.[10] A water right does not imply ownership of the water itself; the right merely approves the use of the water for a beneficial purpose.[11]

There were two authorities or methods in Idaho for acquiring a water right by appropriation—the "statutory" method and the "constitutional" method. The statutory method, expressing procedural details, found its origin in territorial legislation of 1881 and 1887. That portion of the constitution relating to appropriation of water, effective with statehood in 1890, was a more general expression of rights. For example, included was the specific guarantee that "the right to divert and appropriate the unappropriated waters of any natural stream to beneficial uses, shall never be denied."[12] After 1890 the detailed statutory procedures were, of course, required to be in conformance with the constitution. However, any appropriation action could theoretically be based on either statutory or constitutional authority.

Application of the statutory method became prevalent, probably because the statutes contained detailed procedural requirements. Statutes of particular interest to this study were those of 1895, 1899, and 1903, all of which were based largely on the earlier territorial legislation. All these statutes included several primarily procedural provisions associated with appropriation; for example, the posting of notices of appropriation similar to notices of mining claims, construction of the means of diversion, and specific authorization for appropriation of certain waters for storage in reservoirs for future beneficial use. Notices of appropriation were required to be posted in a conspicuous place near the point of diversion of the water from its natural watercourse. Posting and the timely provision of copies to the county recorder and, after 1895, to the state engineer, had the effect of officially and legally making appropriators' intentions known. A broad spectrum of information was required for proper completion of a notice of appropriation, namely:

1. Amount of water claimed, specified in inches in 1887, or in cubic feet per second in 1895, and by later statutes. Initially in irrigation history, the unit of measurement of water flow, acquired from placer mining experience, was the miner's inch. By custom and the statutes of Idaho Territory in 1887, a miner's inch was the amount of water flowing through an inch-square orifice measured under the pressure of a four-inch head. Because of practical difficulties in obtaining such measurements, by 1895 the unit of water flow was changed to second-foot (s.f.)—that flow resulting in one cubic foot of water in one second of time, also termed, in the plural, as cubic feet per second (c.f.s.). The correlation of flow values between these two units of measurement is one c.f.s. equals fifty miner's inches.

2. Beneficial use to which the water was to be applied.

3. Description of the place of intended use. For example, if the purpose was for irrigation, a description of the lands to be irrigated was necessary.

4. Accurate description of the point of diversion. For surveyed lands, description by legal subdivision came to be specified. If the lands were unsurveyed, description by reference to a natural landmark was suggested, provided the location of the point of diversion was described sufficiently "for a person acquainted with the country to find the point from the description in the notice."[13]

5. Description of the means of diversion proposed. Type, size, general course, and length of the canals were features that were to be described.

6. Length of time estimated to complete the diversion works. A maximum of five years was allowed for completion. The requirement to include this item in a notice of appropriation first appeared in the statute of 1895. It was omitted from the 1899 enactment, but appeared again in the 1903 statute.[14]

An aspect of the statutory procedure for appropriation of water—that involving the posting of a notice at the point of diversion and the recording of this posting—was discontinued in 1903. The 1903 statutes substituted a procedure that placed the regulation of water under the jurisdiction of the state engineer.[15] Under the new procedure, those who intended to appropriate water were required to apply to the state engineer for permits before commencing any work on the physical system. A completed application for permit included the same general information as previously required for a posted notice, that is, identification of the applicant, quantity of water claimed, the source of supply, location of the point of diversion, the purpose for which the diverted water would be used, and a description of the proposed works.

The various statutes on appropriation of water also addressed the construction of the means of diversion. Work was to commence within sixty days of the posting of the notice or, after 1903, by the date specified in the state engineer's approval of the application for permit. Work was to be pursued "diligently and uninterruptedly"—except for unavoidable, temporary delays owing to rain, snow, or cold—until completion. As mentioned earlier, completion within five years was the official expectation, although provisions for "departures from normal" existed. For example, under the 1903 statute, there was, as part of the administrative paraphernalia, a special form entitled "Application for Extension of Time for Beneficial Use Proof." Completion was defined as meaning "conducting the waters to the place of intended use."[16]

Some flexibility in the construction as initially visualized was provided. The point of diversion could be changed if no injury to other parties resulted. Canals could be extended beyond the original locality of intended use. A right of "gradual development" was recognized for expensive or long-term projects.

The 1895 statute included provisions for water appropriation in connection with reservoirs. In this instance, the unit of measurement was the acre-foot, which denoted the quantity of stored water covering one acre to a depth of one foot. (One acre-foot equals 43,560 cubic feet of water.)

Under the 1895 statute:

> Any person, association or corporation desiring to construct a reservoir for the purpose of storing water for some beneficial purpose shall have the right to take any of the public waters of this state which are going to waste at any time, or which are unappropriated, and to appropriate and store the same for future beneficial use and to construct and maintain the necessary dams, canals, conduits or other works for impounding and distributing such water, by complying with the same rules and in the same manner as in the Act provided for appropriation and diversion of any of the public water of this state.[17]

As earlier noted, the two methods of appropriation—constitutional and statutory—were of equal validity. Either could "be followed by an intending appropriator at his option."[18] However, application of the doctrine of relation made it further advantageous to adhere to the statutory method when determining completion of the appropriation. Under this doctrine, if the construction requirements, as stated in the notice of appropriation or the application for permit, were met according to the statute, the priority of rights was related back in time to the posting of the notice or the filing of the application for permit.[19] Under the constitutional method, such relation back in time was not possible.

Completion of the procedures of appropriation and of the construction associated therewith did not bestow a clear and uncontestable right to the water. This was true even under the 1903 statute wherein the normal sequence of actions was the application for permit by the appropriator, approval by the state engineer, provision of proof of completion of works by the appropriator, and the granting of a license by the state engineer. The license was "merely a contingent right."[20]

A clear bestowal of right to water had to come from adjudication by the proper court. From the judicial processes would stem an "action to ascertain, determine, and decree the extent and priority of a water right."[21] The decree of a court was to be based on a consideration of the elements of the right. The elements of a water right were three: (1) the priority of appropriation, (2) the extent of the right in quantity, and (3) the extent of the right in time or the period of use for a beneficial purpose.

Both the state constitution and all applicable statutes addressed priority of appropriation. "Priority of appropriation shall give the better right as between those using the water."[22] "Between appropriators, the first in time is the first in right."[23]

The determination of the extent of the right in quantity was more complex. Here several considerations applied. First, the measurement of quantity—after 1895 in cubic feet per second (c.f.s.)—would occur at the point of diversion.[24] Second, during this early period in state history, the extent to which a water claim would eventually be supported by a decreed right was potentially limited by the capacity of the diversion works constructed. Such construction would "have secured this right to the water claimed, to the extent of quantity which said works are capable of conducting, and not exceeding the quantity claimed."[25]

Contemporaneous with this factor of the capacity of the diversion works, and later to supercede it in importance, was a consideration of the real needs of the appropriator.[26] To the requirement of beneficial use, statutory law added the requirement that the appropriator exercise economy and reasonableness of use of water.[27] This concept was called "duty of water." It was "the policy of the laws of Idaho to require the highest and greatest possible duty from the waters of the State in the interest of agriculture and other useful and beneficial purposes."[28] Pertinent to this consideration was the 1887 prohibition against water usage in excess of that required by good husbandry for the crops cultivated.[29] Statutory law did not define the amount of irrigation water usage justified by "good husbandry." However, a later attempt to evaluate the requirements of the land was a step in this direction. The 1903 statute addressed duty of water in this fashion: "No one shall be authorized to divert for irrigation purposes more than one cubic foot of water per second for each 50 acres of land to be irrigated, unless it can be shown to the satisfaction of the State Engineer that a greater amount is necessary."[30] One final point relating to the requirement for strict economy in water usage involved the means of conveyance. An appropriator of water was expected to construct and maintain proper canals and ditches to minimize water loss.[31]

Determination of the extent of a right in time was also important.[32] It was conceivable that water appropriated for a specific beneficial purpose—irrigation, for example—could have only a seasonal applicability. To this end an irrigation season was defined by statute as commencing April 1 (or April 10) and ending October 15 (or October 31), the specific day depending on the statute applicable.[33] Restrictive implications for water appropriated for irrigation purposes are obvious.

The Idaho water law and the policies derived therefrom—evolving through the territorial days and early days of statehood—were crucial to developments within the state. Not only was the law in existence for control of contemporary irrigation and power developments, but the law then existing also provided the framework for future water law and policy.

CHAPTER FOUR

Earliest Irrigation Efforts
in Gem Valley

Whereas in the previous chapter the focus is restricted in time, it is desirable to restrict the focus geographically in this chapter. Attention is limited to the central and southern portions of Gem Valley—the area through which the Bear River flows soon after its turn to the south. Located in this area are the former and present communities of Grace, Turner, Central, Lund, Bancroft, Niter, Bench, and certain agricultural lands south of Central and also west of Lago.

Settlement of the south-flowing arm of Bear River proceeded from south to north for two reasons—proximity to other settled areas as the settlement advanced and the increasing difficulty of obtaining the desirable river bottom lands in Cache Valley and lower Gem Valley. Mound Valley and Cleveland, on the extreme south of the valley, were first settled around 1870. Thatcher was settled in 1881. Lago, Grace, Turner, Bench, and Central were settled in the late 1880s and early 1890s.[1] These latter settlements brought homesteaders to the high, arid tableland adjoining the Bear River as it passed through its deep, rocky gorges. These latter settlements also brought to the valley several of the principal individuals involved in the early irrigation efforts. Among the earliest settlers were Edward J. Turner, John J. Trappett, John Allsop, and Samuel Egbert.[2]

The settlers attempted dry farming with little success.[3] Water was scarce, which made the possibility of irrigation a matter of concern even before the decision to settle the valley. John Trappett illustrated this point. In 1895 "we prospected up the river to see what show there would be of getting the water out before we went on to Blackfoot to file on our homesteads."[4] As could be expected, attempts to divert water from Bear River onto the land received early attention. An early Gem Valley pioneer family history (John Ira Allsop) treats this problem. "They never gave up hope of bringing water on to the land. The water situation was acute. Many of the settlers had to haul their water for

culinary purposes in barrels from Bear River and they felt that without
sufficient water to irrigate the land they would have to give up their
claims."[5]

There is only sketchy evidence of the two earliest efforts at irrigation
in Gem Valley.[6] Involved were Edward J. Turner, Martha Turner, Mi-
chael Emart (possibly Emmett), C. Mickelson, and G. Mickelson. Their
claims were for 100 c.f.s. of water on November 26, 1885, and an
additional 50 c.f.s. on November 26, 1888. Appropriation reportedly
included the posting of written notices and compliance with the laws
"then in force and effect." The point of diversion for both of these
appropriations was stated to be "a point known as the Ten Mile Bridge"
in the immediate vicinity of present-day Grace. The canal associated
with this effort was called, at least in later years, the Turner Canal. By
1900 the only trace of that diversion was a broken-down wooden flume
in the river canyon.[7]

Another diversion was undertaken in 1889 by Samuel W. Egbert,
Hyrum S. Egbert, John Steadman, and Brigham Sellers, and dated to an
appropriation of 86 c.f.s. by the posting of notice on September 2. This
proved to be an entirely different matter as concerned survival of physical evidence of the diversion. This effort, involving a greater investment
of time and labor, was also significant in that the project was begun
almost as soon as Samuel Egbert had arrived in the valley. The point of
diversion in this instance was described as being about at the center of
Section 31, Township 9 South, Range 41 East of the Boise Meridian.[8]
This would place the location about a half mile down river from the
present irrigation dam. Diversion required something in the river to
raise the water level, flumes to convey the water along the rocky canyon
walls, and canals or ditches. The obstruction in the river was not a dam,
but merely "a few cottonwoods snaked across the river" and a "little
brush and rock just below north point of White Tail Mountain."[9] The
flumes were more elaborate. Built of wood and supported by wooden
posts, the flume construction ran along the south side of the canyon bed.
These flumes, unfortunately, were unable to withstand the weight of the
accumulating winter snows and were a major weakness in the performance of the system.[10] The canal itself, built where the soil layer permitted, was relatively large. Trappett described the canal width as perhaps ten feet at the top and eight feet at the bottom.[11]

The Egbert Canal took two years to build but was considered an
unsuccessful, "heart breaking attempt."[12] By 1895 only "tumble down

flumes" and the land tracing of the canal remained. But water had been diverted onto the land. The evidence was clear—plowed and cultivated ground, growing crops, and shade trees that were unable to grow in that country without supplemental moisture. [13]

The failures of the Turner and Egbert canals in the late 1880s and early 1890s were bitter disappointments, but this did not stop the efforts to irrigate. Another water filing, this one involving D. D. Sullivan and F. H. Riddish (possibly Reddish), is noted with the date of April 25, 1898. The point of diversion was described as the southeast corner of Section 30, Township 9 South, Range 41 E.B.M., which would place the location near the present irrigation dam on this section of the Bear River. Three aspects regarding this water claim are unusual. (Indeed, the claim may have been specious.) One is the amazingly large amount of water claimed—6,380 c.f.s. [14] Another is the absence of surviving physical evidence, although this may have been eliminated by later developments in the same area. And finally, this filing was made while other irrigation efforts, begun in 1896, were in progress to eventual success.

In considering these early irrigation efforts, the important points are the high priorities assigned to such activities and the never-say-die determination of the Gem Valley settlers. They "refused to give up. They decided to make another attempt and a dam site was selected one-half mile east of the river bridge. They began hauling logs from the canyon and worked nearly all winter, but for lack of finances, and upon the advice of experts who deemed the site unsuitable, this attempt was abandoned." [15] With the failure of these early irrigation efforts, it is not surprising that the next try would be considered a final effort—a last chance.

CHAPTER FIVE

The "Last Chance"
Irrigation Project

Although the earlier attempts at irrigation in Gem Valley were dis-
couraging, the elements for eventual success were present in 1896. The
activities of an early Grace settler, John J. Trappett, were of crucial
importance. He arrived on his homestead in February, and almost immediately "dam meetings" began to be held "with men from all over the
valley attending. Just what kind of a dam they could afford to build
seemed to be the most important question."[1]

At a meeting in October the settlers "organized a water company to
get water."[2] The initial organization of this as-yet-unnamed company
included John Trappett as president, a position he held until 1898.[3]
Several features relative to this first irrigation company are of interest.
First, the group was small, consisting of only seventeen individuals.
However, future expansion of membership was anticipated. New settlers
"kept dropping in, every few days there would be some one come in
fresh." Second, financial resources were meager—almost nonexistent. As
John Trappett summarized: "We [had] nothing only our homes, that is
all we had." Only Edward J. Turner of the original organization was
considered to be a "man of property."[4] Contribution of work instead of
money was to be the normal means of support.

But these disadvantages did not preclude an energetic approach to the
work to be done. Construction of a dam across the Bear River was
recognized as the "first thing," and by November 1896 the company
was acquiring timbers for the dam.[5] By 1897 the fledgling company had
grown from the initial seventeen to a total of "40 or 50"[6] and had
received a new name. The first filing on Bear River water by this group
occurred early in the year and dam construction proceeded as fast as the
members were able.

When the name "Last Chance" was affixed to this irrigation effort is
not entirely clear. However, the name, obviously appropriate in the eyes
of Gem Valley residents, is generally attributed to John Trappett.[7] In

27

any event, the first water filing for 400 c.f.s. by D. D. Sullivan, John J.
Trappett, and George Stoddard identified those claimants as members of
the Last Chance Irrigation Company.[8]

The exact date of this filing is likewise unclear. John Trappett recalls
the posting of the notice "on the dam . . . about the 14th of February,
if I remember right."[9] However, other dates appeared when a copy of
the notice was presented for notarizing. In this instance posting of the
notice on February 24, 1897, was asserted, which assertion was attested
to by notary public on March 1. The state engineer's office received a
copy of the notice of water right on March 3.[10] These variations in dates
would prove troublesome in the future.

The intended use of the water was specified "for irrigation and culinary and such other purposes as we may desire."[11] Completion of the
water diversion to achieve those purposes was scheduled for "within five
years," consistent with specific provisions of the applicable Idaho statutes of 1895.[12]

Other information required in a posted notice of water appropriation
lacked the precision that would have been preferred by lawyers involved
in future litigation. For example, the location of the point of diversion
was stated to be "near the point where a copy of this notice is posted and
more definitely described as . . . South East of Section 30 Township 9
Range 41 E.B.M. near the North End of what is known as the White
Tail Mountain." Both the section and township descriptions were faulty.
(The correct description would have been the Southeast Quarter of Section 30, Township 9 South, Range 41 East of the Boise Meridian.)

Similarly, the place of intended use is described broadly or incom-
pletely as "Townships 9 and 10 Ranges 39 and 40 E, and Township 10
Range 41 E.B.M." Description of the proposed means of diversion was
also misleading. The means of diversion were specified as "a dam 8 or 10
ft. high and flume about 3000 ft. in length and 2 canals." Each canal
was to be about eight miles in length; one would proceed southeasterly
and the other northwesterly.[13] As will be shown, the Last Chance system
as eventually constructed did not reflect the details earlier outlined in
the notice of water right—and therein were the origins of yet another
problem. (See maps on pages 36 and 39.)

Many specifics of the plans or intentions of the Last Chance officials
were not disclosed by the posting (in late winter of 1896–97) of the
initial notice of water right for 400 c.f.s. of water from the Bear River.
Although details were not announced at the time, twenty-two years later

John Trappett was able to describe with obvious pride a grandiose plan "to water from the Port Neuf [*sic*] to Trout Creek on the south from the beginning."[14] If a plan of such scope were brought to fruition, the Last Chance system would have provided irrigation to lands from Chesterfield, north of Bancroft, south to the vicinity of Lago—almost all of Gem Valley. When pressed for a more precise statement of the plans, John Trappett appeared to be positive that the original intention was to provide irrigation to forty thousand acres.[15] The manner of arriving at this figure was indirect. "We had representatives from each ward and they represented about so many hundreds or thousands of acres of land, how much there was in each ward, that is how we figured about 40,000 the first two years."[16] Shareholders' homestead rights, that is, the total land filed on, became the basis for estimating a total of forty thousand acres.[17] Whether it was topographically possible, in view of ground elevations, to distribute water over all of that land was not considered. However, a meaningful correlation was seen between the forty thousand acres and the appropriation of 400 c.f.s. of water.[18]

Although some vagueness and uncertainty may have existed regarding the exact lands eventually to be irrigated, construction of the means of diversion was not delayed. The dam, a first priority for acquisition of materials as early as fall 1896, continued to demand much of the settlers' time and efforts. Final site selection was also a major consideration, and the settlers sought help from expert surveyors. Several names appear in the record as performing this function—a "Mr. Fryar" who "made the first survey on the east side of the river from Alexander Point;"[19] a "Mr. Haliday" who was said to have used a gun barrel for a spirit level in making the "first survey;" and a "Mr. Atkinson" from Richmond, Utah.[20] The site finally selected for the dam was located about a mile and a half downstream from Alexander Point,[21] about the same location on the river as had been the "old Egbert dam"—an earlier diversion effort.[22]

A dam using locally available materials—timber and rocks—was visualized. The first step was to build log "cribs," position them in winter on the surface ice over the width of the river, fill them with lava rocks, and wait for the melting of the ice in the spring and the settling of the cribs to the bottom. Each crib was about thirty feet wide.[23]

Once the cribs were in place on the river bottom to form the main foundation of the dam, the level of obstruction to the river flow could be raised—and thus a dam was built. The settlers hauled logs for this purpose from nearby canyons. Also, some logging was done from the mountains bordering the river to the north, the logs being floated downstream to the dam site.[24] These were large timbers, some reaching sixty feet in length.[25] To submerge these logs, boulders were dislodged and rolled down the hills bordering the dam site and placed on top of the logs. To one participant in the construction, some of these boulders were seen "as big as a load of hay."[26]

As the dam rose, it became clear that additional height would be required to accomplish the water diversion as planned. Whereas the 1897 notice of water appropriation had visualized a dam eight or ten feet high—thus making prior state approval of the dam construction unnecessary—permission from the state engineer was now required for a height more than ten feet.[27] Approval was obtained and the dam was raised an additional two to three feet.[28] Cross ties were placed on the top, and timbers were placed with the current to form a "directional channel" to improve water flow.

By the winter of 1897–98 the dam had been completed to a width of 190 feet.[29] A "loop hole in the crib work," left purposely for the river to flow through, was finally closed after two years. Thereafter, the water spilled over the top of the dam at the west end.[30]

Construction of the dam was a major engineering challenge. Success in the effort was an essential prerequisite to later construction of flumes and canals and eventually bringing the water onto the land. Particularly significant was that this rough, dangerous work was accomplished with no modern construction machinery or equipment. The workers used farm tools typical of the day, such as crow bars, shovels, picks, and other crude hand instruments.[31]

Additionally, the project was hampered by nature, shortage of money, wide dispersal of the work force, and extremes of weather and climate. For example, on the first day of the site selection survey, the crew killed fifty-nine rattlesnakes in the immediate area.[32] Money was so short that adequate clothing was often not available. Trappett noted, "we couldn't get a pair of overalls, we had to use two-bushel sacks to make our overalls to do that work with."[33] Another individual, close to the work through family connections, wrote of workers "coming for miles with horses, often poorly clothed, feet wrapped in burlap against freezing."[34] Inadequate food drew the attention of others. Food was typically limited to bread and gravy, supplemented perhaps by a bit of bacon for breakfast. Lunches often consisted only of bread and syrup. Sharing of lunches and other examples of cooperation were common features. "Many times after work the men had to stop at the Sullivan home nearby to thaw the wet and frozen burlap from their feet and get warm before going home."[35]

By the spring of 1898 the water behind the dam was high enough that attention could be shifted to the means of moving the water from the point of diversion onto the land.[36] Those means included both canals, where the gradient and soil conditions permitted, and flumes.

Construction of canals involved much blasting and scraping to establish the bed.[37] Teams of horses with scrapers and men with dynamite and picks were visible features of the project. The initial flumes were made of wooden planks supported by wooden posts. A typical flume was ten feet wide and six feet deep. Supporting posts were high enough so that an additional foot of depth could be achieved by adding other planks if an increase in flume capacity was needed.[38]

The initial plan was for an open canal south from the dam as far as possible along the west bank of the river.[39] Flumes would be used only where canals were impractical but would, of course, be necessary to cross the Bear River canyon. As the work proceeded on the "inter-mix" of canals and flumes, many adjustments in the details of the initial plan became necessary. For example, diverted water immediately spilling from the dam was so swift that it washed out the open canal, necessitating replacement by a section of flume 250 feet long.[40] All this difficult work over two hard years notwithstanding, no farmland had as yet been irrigated.

While construction was still in progress, other changes affecting the "Last Chance" occurred. On February 14, 1899, the community-oriented irrigation effort was formally incorporated as the Last Chance Canal Company (Ltd.), with a growing list of sixty-four original stockholders. Edward J. Turner was the first president. Fifty thousand shares of stock were issued at one dollar per share.[41] Only 12,272 shares were subscribed at the time of incorporation, however. Sale of the remainder, or 75 percent of the total, would have to await procurement of more funds—or the arrival of new stockholders.[42] Money was so scarce that even the initial stock issues—and later company assessments—were paid with credits (at $1.50 per day) for work done or pledged.[43] Although details of any transactions are lacking, apparently there was also some local trading and selling of Last Chance stock.[44]

Another significant event, not directly related to the ongoing construction, was a second water filing that took place in May 1901. This filing for 600 c.f.s. of Bear River water, initiated by Edward J. Turner as president of the Last Chance Canal Company, was legally more satisfying than the initial 1897 filing. But even in this instance some confusion on the date was present. The notice of water appropriation was indicated to be May 11, 1901; however, the document was not notarized until May 14, and that same date was shown for the posting of the notice. The proposed use of the water appropriation was specifically for "agricultural and domestic purposes." Gone was the additional overly broad 1897 expression of "such other purposes as we may desire." The point of diversion was much more precisely defined than in the initial filing and was further identified as a point where "a dam and head-gate have been constructed by the said company." The means of diversion were described as the "dam already constructed" and a ditch and branches "having a total aggregate length of about one hundred (100) miles." The places of intended use, following the legal subdivision descriptions, were summarized as containing in all about seventy-five thousand acres.[45] The company committed to complete the diversion within five years, even though the applicable Idaho law no longer contained this specific requirement. Now the law required only that the work for diversion commence within sixty days of the filing and that it proceed "diligently and uninterruptedly to completion."[46]

There were several reasons for the 1901 filing for 600 c.f.s. of water. An obvious reason was the increase to seventy-five thousand acres in the acreage intended to be irrigated. The increase was at least partially to accommodate individuals who would file on land after the irrigation project was completed and thus give the Last Chance some capability for expansion.[47] A suggestion of intent to sell water from this additional filing was vigorously denied.[48] Other considerations suggested an insufficiency in the amount of water appropriated in 1897. One dealt with aspects of soil and climate. It was "discovered that it took more water to water that land than a person would naturally think."[49] Further, an effort was made to relate shares of Last Chance stock to acres to be irrigated and to inches of water available. It was found that "none of us would have sufficient water to water the ground."[50] However, the primary reason for the insufficiency was probably an error in computation made when determining the amount of water represented by the 1897 filing. Last Chance officials had "figured on 144 inches to the foot,"[51] rather than the correct measurement equivalent for flowing water of one c.f.s. equaling fifty miner's inches. Thus, the original filing (1897) represented only 35 percent of the amount of water believed to have been appropriated. Regardless of the accuracy of the computations, however, it appears that the Last Chance requirements for water appropriated—totaling 1,000 c.f.s. by May 1901—were only broad and rough estimates.

In 1919 lawyers pressed the indomitable John Trappett to explain why the various Last Chance water appropriations were made. His reactions were characteristic and enlightening. Concerning lapses of memory about intentions, he complained that "nobody ever thought anything would come up years afterwards and no one would ever think of it."[52] When the suggestion was made that the water appropriations may have been excessive, his response was: "We should have filed on all of Bear River."[53] And then for any mathematical shortcomings or inconsistencies, he somewhat disingenuously absolved himself of responsibility with the observation: "I am no figurer myself."[54]

As the end of the initial five-year construction period approached, it became obvious that there remained several problems requiring resolution. First was the matter of time itself. Based on evaluation by Last Chance officials, it appeared that "we couldn't make it in time, so we bonded to William Slick to put this flume in."[55] The flume in question was the one that received water from the original open-ditch feeder system (specifically the Last Chance Main Canal) which coursed south from the dam on the west side of the river and then crossed the "river at the bend in the river."[56]

But the dam itself turned out to be the worst problem of all. In its construction, adequate provisions for the headgates had not been made. The gates had been left until the "last moment."[57] When construction on the gates finally began it was found that the structure of the dam itself made installation of the headgates extremely difficult. The gates were to be inserted in the west end of the dam—unfortunately already the location of especially large and well-embedded timbers and boulders. "It would have been an awful job to complete and fix up that dam . . . and sink our gates in through those big logs."[58] And, as if marveling at what had developed, John Trappett, speaking almost twenty years after the event, said: "I don't know how it occurred, but it occurred some way, that dam we already had in, he [Bill Slick] persuaded them to put in a dam just below, probably forty feet . . . below the dam we had put in."[59]

The decision to construct a new dam involved selecting a contractor and obtaining funds. The firm J. B. Slick of Salt Lake City became the contractor. The company borrowed fifteen thousand dollars from Miller and Viele, a loan company in Logan, Utah. Security for the loan required clarification of potential water rights. It was discovered that the initial filing of 1897 for 400 c.f.s. of water accrued no benefit to the Last Chance Canal Company.[60] That filing, it will be recalled, was by John J. Trappett, David Sullivan, and George Stoddard, as stockholders of the Last Chance *Irrigation* Company. A quit claim deed of October 22, 1901, wherein note was made that the Last Chance Canal Company (Ltd.) had been "improperly styled . . . The Last Chance Irrigation Company" in the 1897 filing, and that conveyed all rights under the 1897 filing to the Last Chance Canal Company (Ltd.), was executed by Trappett and Sullivan.[61]

Bill Slick, who had been so instrumental in contracting for the flume and second dam, was regarded with admiration by Gem Valley settlers, and both the flume and the dam came to bear his name. John Trappett, a handy man with a well-turned phrase, having named the Last Chance system, provided his evaluation. "We wanted it [the dam] to have that name, calling it Slick all the time, because he was pretty slick and we named it particularly for him."[62] Unfortunately, information about construction of the Slick dam is not available.

The original dam, built at such great costs in time and effort, was never completed; it never diverted any water from the river.[63] But Trappett remained almost aggressively protective of the worth of that project. Under questioning, he asserted that "it could have been just the same as the other dam."[64] In 1918 the old dam was still in place.[65]

All of the money was gone by early 1902. Another loan, this one for five thousand dollars, was obtained from Miller and Viele on February 3,[66] and construction continued. An awareness of the five-year maximum time allowable by law for conducting the water to the place of intended use was always present. This period expired on February 14, or February 24, or March 1 of 1902, depending on the date assumed for the first water appropriation filing. But regardless of the exact date, the Last Chance people were continually concerned with having the construction completed "within the time of having the water out, that is all we were figuring on."[67]

Uncertainties regarding conclusion of the first phase of the project brought an instance of high drama. Two weeks before the time expired for completion, the Last Chance workers still had 1,100 feet to go on the Last Chance Main Canal system and the ground was frozen so hard it was impossible to work it. A family history picks up the story:

> Mr. Slick . . . said it was impossible to finish the ditch in the allotted time. He said the Bothwell Canal Company was ready to file on the water if they didn't have it finished by the twelfth.
>
> Everyone had left the works feeling downhearted and discouraged, with the exception of Slick and his helpers, and Frank Christensen. Christensen after much thought, suggested that they make a new ditch. Slick thought he was out of his mind from worry, but told him to go ahead. He got some of the men to come back and they plowed and scraped an eleven hundred foot ditch through the snow. This thawed the ground and they were able to build the flume. The ditch was finished two days before the deadline and the filing was saved.[68]

Thus, by February 1902 "Last Chance water" was available to farms on the east side of the Bear River. Irrigation was underway.[69] Another flume across the Bear River further downstream was constructed in

The Last Chance Diversion Dam
on the Bear River (1982)

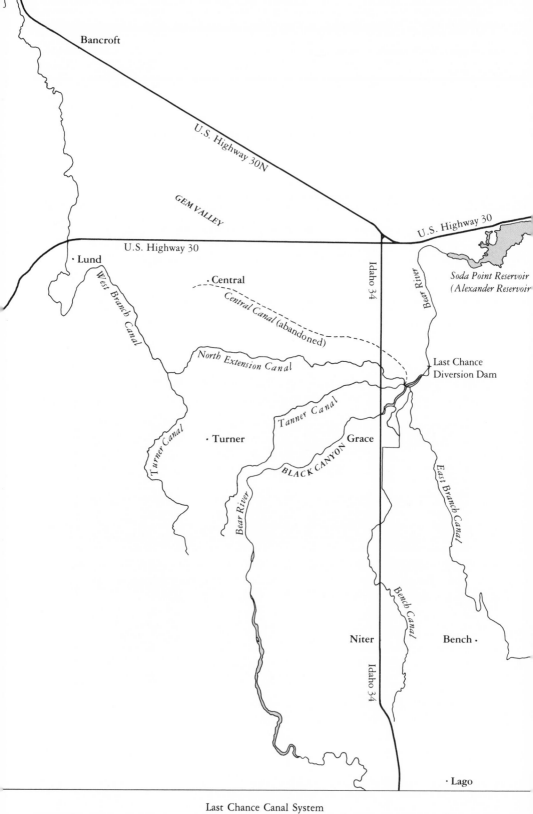

Last Chance Canal System

March and April 1902 to bring water back to the west side.[70] All this initial flume and ditch work from one side of the river to the other had finally "succeeded in getting the water onto the land far enough to legally comply with the stipulation of the law."[71]

The significance of these accomplishments was not lost on the local residents. A feeling of exhilaration prevailed, as exemplified by the following account.

> As a little girl I [Alice T. Sorenson] can remember when the last work was finished on the last canal and my father (George Telford) along with the others came home standing up in their wagons swinging the lines around their heads and shouting and singing with the joy of it. Wagons could be heard rumbling all over the valley for some time that evening.[72]

The Last Chance Canal as finally constructed bore only slight resemblance to the means of diversion described in the notices of appropriation of 1897 and 1901. In those two documents, flumes 3,000 feet in length and two principal canals, each eight miles long, and a total length including branches of about one hundred miles, were visualized apart from the dam. What had actually been built to convey the diverted water from the headgates at the new "Slick" dam were a flume 274 feet long immediately south of the west end of the dam, 1,100 feet of open ditch, 1,800 feet of flume (the Bill Slick flume, which extended south to and across the River to the east side), and another 1,800 feet of open ditch to headgates of two east-side canals. From this point another flume conveyed water back across the river to the west-side canals. The tortuous path of the combination of open ditches and flumes was dictated by the "line of levels" survey starting at a diversion dam of optimum height and running through irregular and rough terrain. The total length of the works constructed and owned by the Last Chance Canal Company (Ltd.) was about one and one-fourth miles.[73] The capacity of this portion of the works in 1902 was 450 c.f.s.[74]

In place of a single, integrated Last Chance Canal Company, what evolved, as viewed by the Last Chance officials themselves, was a "Last Chance Canal system"—a single and interdependent system consisting of the Last Chance Dam, the Last Chance Main Canal (which served as a feeder), and five primary irrigation canals, each of the latter managed, controlled, and operated by a secondary corporation. Fed directly from the Last Chance Canal headgates were the East Branch and Bench canals on the east side of the river, and the Central, Tanner, and North Extension canals on the west side. The North Extension Canal fed two large branch canals—the West Branch and the Turner.[75] The East Branch Canal, in February 1902, was the first to receive water from the Last Chance.[76] The other primary irrigation canals received water soon thereafter—for example, the Bench Canal in May and the North Extension in July 1902. With the provision of water to the North Extension Canal, irrigation became possible on the west side of the Bear River. As could have been expected, the branch canals were enlarged and extended as rapidly as possible. Within two years all had essentially reached their planned lengths.[77] The 1903 growing season found an estimated 3,500 acres of Gem Valley under irrigation.[78]

Irrigation of only 3,500 acres was a modest achievement when evaluated against the total acreage planned to be irrigated or the Last Chance filings for Bear River water. However, at this stage the degree of project completion was not crucial to later efforts to determine the extent in quantity of the Last Chance water appropriations. Idaho water law recognized a concept of "gradual development" of a project's physical facilities.

Several major problems were associated with the first years of operation of the Last Chance system. The original decision to use wooden flumes in the Last Chance Main Canal was the source of the most recurrent and pressing of these problems for two principal reasons. First, the flumes significantly restricted system capacity. This problem was addressed the first summer of use by the addition of an eight-inch plank at the top of the flume support posts.[79] Further major enlargements of the canal from the end of the flume on the east side of the river were made in the years 1905 to 1907.[80]

But even more objectionable than system capacity limitations were the wooden flumes themselves. They leaked considerably, were subject to decay, were easily crushed by winter snow, and needed constant repair or rebuilding.[81] In 1909 a replacement wooden flume was constructed,[82] but by 1915 the flume problem was again of crisis proportions. John Trappett described the canal as being in "bad shape. . . . We were looking for it to go down all the time, three or four bents would go down all unawares to us."[83]

Such a serious problem justified an innovative solution. The possibility of tunneling through a knoll of solid lava rock west of the offending flumes had been discussed for the "last ten or eleven years, but the

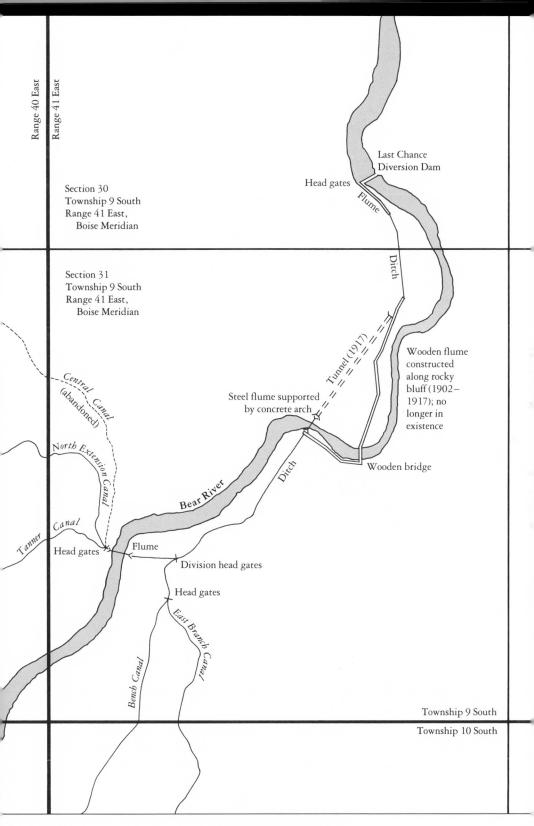

Range 40 East

Range 41 East

Section 30
Township 9 South
Range 41 East,
Boise Meridian

Section 31
Township 9 South
Range 41 East,
Boise Meridian

Last Chance
Diversion Dam

Head gates

Flume

Ditch

Tunnel (1917)

Wooden flume
constructed
along rocky
bluff (1902–
1917); no
longer in
existence

Steel flume supported
by concrete arch

Wooden bridge

Central Canal
(abandoned)

North Extension Canal

Bear River

Ditch

Tanner Canal

Head gates

Flume

Division head gates

Head gates

Bench Canal

East Branch Canal

Township 9 South

Township 10 South

Last Chance Main Canal

people thought it was too big an undertaking." Trappett himself favored efforts "to patch the old flume up . . . and use it just as long as we could." But it eventually became obvious that either a new flume of improved construction or a tunnel was necessary.[84]

On August 7, 1916, Last Chance officials voted for the tunnel option and requested their engineer, William A. Samms, to advertise for bids on the project.[85] With a bid 31 percent under the next lowest bidder, Morrison-Knudsen of Boise was awarded the tunnel excavation contract on September 12 for $16.95 per linear foot. Last Chance expected to pay for this work by assessing ten cents per share of company stock,[86] and by obtaining a mortgage loan on Last Chance property, eventually arranged with Utah Mortgage Loan Company of Logan, Utah, for twenty-five thousand dollars.[87] The tunnel, twelve feet wide and nine feet high, commenced near the lower end of the open ditch bringing water from the diversion dam, ran straight in a generally southwest direction, and met the river about six hundred feet downstream from the then existing wooden flume trestle.[88]

Construction was difficult. The south half of the tunnel was "hand drilled and the rock hauled out with one horse pulling a small rail car filled and dumped by hand."[89] Payments of $23,687.46 were made to Morrison-Knudsen for completion of the contract in August 1917.[90] At the bid-cost of $16.95 per linear foot, these payments cover excavation of just under fourteen hundred feet of tunnel. However, inspection of current maps suggests a tunnel length of just over twelve hundred feet.

The location of the tunnel outlet required the construction of a new cross-river flume. This matter was first considered in August 1916, at which time the company discussed "plans for the crossing of the River with our flumes" and possible alternatives of concrete and steel flumes.[91] By June 4, 1917, Last Chance decided in favor of a steel flume structure seventy-three feet long, with the balance of the flume to be constructed of wood.[92] The steel-wood flume crossing the Bear River was supported by a concrete trestle in the form of a massive arch. Thus, after fifteen years of problems related to the extensive wooden flumes, solution was partially adequate, if not complete.

Still other problems, although minor by comparison, confronted the Last Chance. During 1909 and 1910 additional water filings were made by the Bench Canal Company and the Tanner Canal Company, two "secondary corporations" of the Last Chance "system." The Bench Canal Company, using the "application for permit to appropriate water" tech-

Pre-1917 Last Chance Flumes
on the West Side of the Bear River

nique in effect under Idaho law since 1903, filed for 138.16 c.f.s. and 25.6 c.f.s. of water on August 9 and December 31, 1909.[93] Similarly, on July 29, 1910, the Tanner Canal Company initiated the appropriation process for 54.00 c.f.s.[94]

Several aspects of the Bench and Tanner water appropriation actions are of interest. First, Bench and Tanner company officials selected different points of water diversion than Last Chance. Two points of diversion, one from the east side of the river for the Bench Canal and one from the west side for the Tanner Canal, were located slightly over two miles downstream from the Last Chance dam, and they coincidentally benefited from the nearby hydroelectric dam constructed by Utah Power and Light just north of Grace in 1908. The change in points of diversion for the "B" canals avoided both full reliance on the Last Chance flume system and the necessity to construct their own diversion dams. Second, the new points of diversion required the construction of new, short ditches—the Bench "B" and Tanner "B" canals—to connect to the existing Bench and Tanner canals. Third, the descriptions of the irrigation works and the lands to be irrigated contained in the applications for permit appeared to duplicate the facilities and capabilities already in place with the completion of the Last Chance system. For example, under the 1909 plan, as amended,[95] the Bench Canal was described as 27½ miles of main canal and laterals to irrigate 7,508 acres. The Tanner Canal data showed a conduit 12¾ miles long to irrigate 5,181 acres.[96]

Construction on the Bench Canal facilities was completed on August 22, 1914; the Tanner work was finished on September 23, 1915. Existing law permitted an additional four years to provide proof of complete application of the water to beneficial use. The Tanner portion was certified as complete on September 24, 1919. The original completion date for the Bench Canal—October 8, 1918—was extended one year owing to "scarcity of farm labor, occasioned by taking many men into the military service of the United States."[97]

Although documented evidence is lacking, it is interesting to speculate about possible motivations for the additional water filings of 1909 and 1910 by the Bench and the Tanner canal companies. New points of diversion and segments of new additional canals (the "B" canals) avoided both the system capacity restrictions and the tremendous reliability and maintenance problems associated with the Last Chance wooden flumes—problems not yet solved at that time and fresh in mind from the flume replacement project of 1909. The Bench Canal, in particular, increased

The Last Chance System Tunnel Exit,
Arch, and Flume across the Bear River (1977)

system capacity almost 40 percent.[98] But possibly even more significant, the new filings by the Bench and Tanner canal companies would serve to establish "entirely independent rights"[99] and, perhaps, to relieve any existing doubts about the legal sufficiencies of the Last Chance filings a decade earlier. Further, the additional filings would, if all were favorably adjudicated, supplement the potential water rights of the Last Chance system to a total of 1,212.78 c.f.s. Tending to confirm an aspect of this final speculation, on May 10, 1917, the Last Chance Canal Company took "over the works, water rights and holdings of the Bench B and Tanner B canals."[100] Payment took the form of forty-five hundred shares of Last Chance capital stock, allocated between the Bench "B" and Tanner "B" in proportion to their water filings of 1909 and 1910.

One final event of the early Last Chance years remains prominent in the folklore of Gem Valley. The farmers building the Last Chance had prided themselves on their independence; only when it became essential for completion of the diversion dam was indebtedness incurred. The company completed payment of the bonded indebtedness in 1914, and on February 23 a huge community celebration was held to "honor the

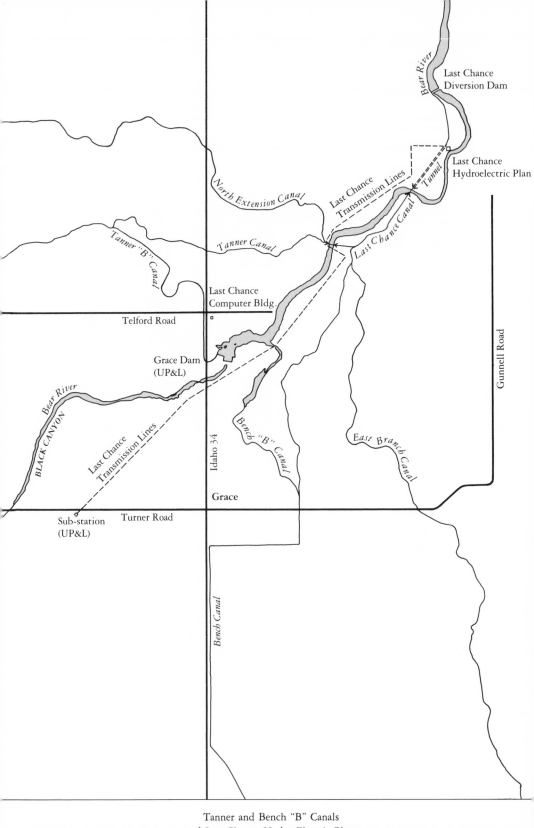

Tanner and Bench "B" Canals
and Last Chance Hydro Electric Plant

pioneers and promoters of the Last Chance Canal Company" and to celebrate "the payment of their bonded indebtedness." The celebration, involving a banquet, speeches, vocal and instrumental music, and dancing, was held in Grace at Columbia Hall, the community social center. The hall was tastefully decorated for the occasion, complete with a sign extending over the stage reading "FREE FROM BONDAGE"[101]—doubtlessly a case of "pun intended."

The capabilities and the effects of the Last Chance system can be clarified by three elements of data—the geographic extent of the system, the canal capacity in cubic feet per second, and the number of acres irrigated. When considered as a whole, by 1919 the Last Chance system consisted of "about 90 miles of main canals and 127¾ miles of laterals."[102] Over the first two decades of its operation, the capacity of the Last Chance main canal fluctuated from 450 c.f.s. in 1902, to 520 c.f.s. in 1905, and down to 428.5 c.f.s. in 1918. The decrease in capacity resulted from a failure to maintain the grade in the canal.[103] Various figures have been recorded for the increase in the irrigated acreage by 1918 from the estimated 3,500 acres in 1903. In 1919 the Last Chance organization itself stated a beneficial use of appropriated water for irrigation of "43,831.3 acres of land or such part thereof as the said quantity of water sufficed to irrigate."[104] Even considering the latter qualification, the 43,831.3-acre figure appears high. A total of 29,000 acres to cover the places of intended use of appropriated waters by the Last Chance Canal Company—including all seven of the "secondary corporations"—appears to be more realistic[105] and was applicable in 1917. A potential discrepancy between places of intended use and acreage actually irrigated must be acknowledged.

The company made periodic improvements to the physical facilities as necessary. The years 1926 through 1929 were devoted largely to improvement or replacement of the remaining wooden flumes.[106] As a commentary on the scope of such work, a local writer concluded that during this period all wooden flumes were replaced by steel ones.[107] However, nearly twenty years later, in November 1945, the matter of flumes became a problem again. Last Chance authorized a study to consider repair or replacement of the flumes (and their substructure) which provided water passage across the Bear River from the tunnel outlet. This study showed a need for a new steel flume 740 feet in length. Installation was completed in early August 1948 at a cost of over fifty-nine thousand dollars.[108]

This long, new steel flume across the river was built above the existing concrete arch and, "although no longer necessary, the arch has never been removed."[109] The decision to retain the "old arch" was apparently because of sentiment or because of a desire to avoid the costs of demolition; the vote within the canal company board of directors was three to two to retain the functionally useless structure.[110]

Over the years, other, less spectacular improvements were also made. The tunnel required smoothing of a high point at the entrance and removal of loose rock and other debris that slowed the water flow and produced moss.[111] Moss growth was a recurrent problem throughout the system. Suppressant chemicals were applied to the Last Chance Canal and to the upper ends of each of the secondary canals,[112] but the results were not recorded.

It has consistently been the objective of the Last Chance system to maintain and to utilize effectively the existing physical irrigation facilities. Optimum use of facilities was shown, for example, by the policy decision for maximum use of the Bench "B" Canal to relieve pressure on the Last Chance upper flume.[113] However, certain problems arose over the years that were major frustrations or that were addressed by eliminating some of the offending facilities. As previously stated, the Bench "B" and Tanner "B" canals were fed by water diverted from the "pond" produced by the hydroelectric dam on the river at Grace. Control of the water level behind the dam apparently was entirely the prerogative of UP&L. If the water level dropped, adverse fluctuations in the water for the "B" canals occurred and the power company was blamed.[114] This problem existed for over twenty-five years. At one meeting of the Last Chance Canal Company board of directors the question was asked, but left unanswered, whether it was not the responsibility of the power company "to deliver our full amount of water."[115]

Experience over an extended time suggested the advisability of eliminating two of the secondary canals from the Last Chance system. The Central Canal passed over terrain of relatively thin top soil and recent volcanic activity. Water losses were great into lava crevices that could not be sealed, and in the early 1930s the Central Canal was eliminated.[116] Only at the Last Chance Main Canal headgates is there any visible indication that a now missing canal once existed.

The Tanner "B" Canal, covering a distance of about one mile from the hydroelectric dam to the junction of the Tanner "B" with the Tanner Canal, presented a variety of problems. Reduced water flow owing to drops in the river level has already been mentioned. Additionally, it was difficult to keep the canal channel clean and the canal came to be considered in "bad shape." Further, there were complaints that seepage from the canal was contaminating nearby culinary water supplies. By late 1969 it was clear that either the canal receive major repairs or be closed. After a two-year delay in selecting one of these two possibilities, in November 1971 the landowner through whose property the Tanner "B" Canal passed received permission to fill in the canal. The Last Chance Canal Company was careful, however, to protect an option for future restoration of the Tanner "B" by retention of "ditch rights" through the fields and by insistence one year later that a thirty-six inch culvert be retained by the state of Idaho where the abandoned canal passed under Highway 34.[117]

One problem that occasionally confronted the Last Chance system involved a matter over which the irrigators had no control. When the natural flow of the river was low, water diversion for irrigation fully according to water appropriations was no longer possible. Under these circumstances, irrigation could proceed only if additional water was pumped from Bear Lake into the Bear River. This was done by "renting" water from UP&L, which controlled the pumping on Bear Lake. The costs of renting this water were of constant concern. The unit of measurement for assessing costs was "per acre foot," and the records show continuous increases. In 1919, the first year that water was rented, cost was $1.15 per acre-foot. By 1931 the increase was modest (15 cents), but an inflationary surge upped the cost to $4.36 in 1976. It was estimated that that would double in 1977. In other terms, in 1920 water rent charges were about $18,000. By 1966 the costs had more than doubled to nearly $38,000.[118]

A matter of organizational importance to the Last Chance Canal Company occurred in 1949. The company was originally incorporated on February 17, 1899, for a period of fifty years, and fifty years had now passed. Action was then taken toward "perpetuating the corporation."[119]

As the years passed, the Last Chance system increased in importance to Gem Valley. In 1968 the Last Chance Canal was shown to have a capacity of 620 c.f.s.—a 45 percent increase over the 1918 data. Acres under irrigation were also up from 29,000 to 33,000 for the same period, an increase of 13.8 percent.[120] However, a more recent and perhaps more accurate determination of the number of acres irrigated by the Last Chance system was 32,000.[121]

TABLE ONE

Summary of Development of Last Chance Irrigation System

Year	Bear River Water Claimed (c.f.s.)	Decreed (c.f.s.)	System Capacity (c.f.s.)	Irrigation Planning Goals (acres)	Actually Irrigated (acres)
1896				40,000	
1897	400				
1901	600			75,000	
1902			450		
1903					3,500
1905			520		
1909	163.76				
1910	54				
1917					29,000
1918			428.5		
1919				43,831.3	
1920		657.76*			
1968			620†		33,000
1977					32,000

*Includes 440 c.f.s. of Last Chance filings and Bench and Tanner filings of 1909 and 1910.
†Includes the "B" canals.

The 1968 report showed 10,720 acres being sprinkle irrigated. It is not clear whether this number is included in the total "acres under irrigation," but from the structure of the report, that was probably the case. In any event, sprinkle irrigation has become a major development of Last Chance resources. This method of irrigation was first used experimentally to place water on areas not accessible to flood irrigation in Gem Valley in the mid-1930s, and, upon proving itself, its use increased from then on. In the decade 1958–68, acreage receiving sprinkle irrigation in Caribou County increased nearly 700 percent; corresponding increases would be applicable to the area served by the Last Chance system, although specific data to support this observation are not available. The attractiveness of sprinkle irrigation was that its users would be "using the same amount of water on more acres, more effectively."[122]

The history of the Last Chance Canal system is largely told through events concerning the construction, operation, and maintenance of the system. However, there is another aspect deserving attention—and this is, in many respects, a "history" of events that did not happen. Even though some matters that received the attention of the Last Chance board of directors were not implemented, their presence in the records of the directors' meetings is clear evidence of consistently aggressive, innovative, and dedicated attitudes. There are many examples, a few of which are cited below.

By late 1920 the value of the influence and prestige coming from power developments on the Bear River had been brought dramatically to the attention of the Last Chance officials, who decided to investigate filings for power rights and to change the Last Chance Canal Company articles of incorporation to permit power-related activities. A power site was apparently selected and appropriate filings made. However, in 1925 the canal company reached a decision to dispose of both the site and the filings. Last Chance's interest to become involved in generating electrical power apparently laid dormant for over fifty years, at which time the possibility of a Last Chance hydroelectric company was again raised. [123]

General water availability or adequacy was always a concern. Continued availability of water in the canals for culinary purposes during the "non-irrigation" season was important, especially in the early days. "Jousting" with UP&L over water adequacy was a recurring subject that took several forms. UP&L had bought stock in the Last Chance Canal Company, and Last Chance officials asserted that the power company was using their stock to divert water from irrigation needs to the generation of power. As another example, in 1929 inclement weather delayed the opening of the irrigation system beyond the normal date in early April. Retroactive recovery of the thus-unused water that had gone into power company reservoirs was suggested. A similar desire to reclaim water, but based on a different logic, was expressed in 1956. In this instance the argument was stated: "If it could be established that the water leaking from the Last Chance Canal was being returned to the River which parallels the canal that we request the Power Company to restore the water to the Last Chance Canal Company." Another intriguing possibility concerned water from springs beneath nearby Soda Springs Reservoir. The Last Chance system had been receiving credit for 4 c.f.s. of water from these springs. To exploit the low water level in the reservoir in 1979 and to satisfy a suspicion that the credit allowed for the spring flow had been inadequate, the canal company proposed close examination of the springs to verify "that there might be more than that there."[124]

Water rental was treated differently in 1972. Last Chance defined an initial objective of establishing a "user's right" on a permanent basis to the water previously rented over the past half century. This was later modified slightly to recording water use by Last Chance over the years to discourage any "others who could be looking at this water."[125]

As has been illustrated, Last Chance officials jealously guarded the company's water rights against all sources of competition. To this end a company policy was formally stated of "protesting and contesting any water application which could conceivably affect the natural flow of Bear River, and thus the company's water right." Application of this policy ultimately resulted in a strange alliance, that of Last Chance and UP&L, in opposing all attempts to vary adjudicated water rights or priorities.[126]

The canal company board of directors in the mid-1970s was oriented toward investigative examination of two rather visionary and large-scale modifications to their irrigation system. One of these possible projects involved converting the entire canal system to a "gravity flow pipe system." A rough estimate of forty million dollars for such a venture put an end to this consideration.[127] The other possible project was directed toward acquiring a dam site at Soda Springs to store water for future irrigation use.[128] A feasibility study in June 1981 by the Idaho Water Resource Board "considers constructing a dam on the Bear River near the City of Soda Springs, Idaho . . . and examines consumptive use of water for supplemental irrigation and thermal power plant cooling."[129] Needs for water had again made "strange bed-fellows." Last Chance was interested in this dam and reservoir to provide a supplemental source of water for irrigation; the power company was interested in the same facilities to provide water for cooling a proposed 2,000 megawatt coal-fired electrical plant, perhaps to be established in the vicinity.[130] This possible project remains dormant.

Thus ends discussion of almost a century of the history of the Last Chance system, except for several specialized aspects covered in chapters eight and nine that stem directly from hydroelectric developments. The stories of the Last Chance as an irrigation system, highlighting the hardships, the drama, and the essential nature and magnitude of the project, make up much of the legend and folklore of Gem Valley. "It can be said of the men who organized this company and carried on the

Memorial Plaque, Last Chance Canal Company,
at the City Park, Grace, Idaho

work necessary to complete the same, that they were real pioneers in
every sense of the word. They were willing to make the sacrifice necessary without a murmur of discouragement always for going ahead and
helping one another in common endeavor."[131] Historical recognition of
the "tremendous task . . . accomplished through the ingenuity and per-
severence [sic] of the brave men and women who pioneered this valley"

was accorded by a plaque and monument erected in 1955 by the Daughters of Utah Pioneers in the city park at Grace, Idaho.

It is difficult to exaggerate the importance of irrigation to Gem Valley. The effects of the Last Chance on the lives of the people indeed have been dramatic.

CHAPTER SIX

Reservoir and
Power Developments

In the American West in the late nineteenth century there was an acute need for a cheap and flexible form of power in the mining and smelting industries. In the earliest developments toward mechanization of these industries, steam power had been used, with the steam produced from wood fuel. This continued until the hillsides were denuded of trees. Later, expensive coal was hauled to the mining or smelting sites, usually by burros.[1] A replacement means of power was sorely needed in these industries.

Advances in hydroelectric power generation showed promise. Essential elements for production of hydroelectric power included a stream with adequate water flow and fall, a dam, and a reservoir above the dam in which to impound water to insure a constant flow during the natural decline of water volume during the late summer season. Electric power was produced by applying the force of the stream to a water wheel connected to an electric generator. All the early hydroelectric plants followed this concept.[2] In addition to the essential elements, canyon formation on at least a portion of the course of the stream was desirable for economy in dam construction along with a location in which to store the reservoir water without flooding extensive areas suitable for agriculture.

The Bear River watershed provides almost optimum conditions for hydroelectric development. The drainage area is large. The river is almost five hundred miles in length, with numerous narrow canyons and a fall of over fourteen hundred feet over the canyon sections of the river. And finally, Bear Lake, near the upper reaches of the river, provides a natural reservoir.

Competition for these valued resources would be great between those zealots in the Reclamation Service who saw irrigation as the prime consideration and private interests that favored hydroelectric development. Control of Bear Lake appeared to be the key to effective use of the entire watershed area.

Federal government involvement in Bear Lake came first. In 1888, as an extension to his 1879 survey of public lands, aridity, and reservoir sites, John Wesley Powell, director of the U.S. Geological Survey, was successful in obtaining legislation that authorized the Geological Survey "to identify and segregate the most promising reservoir sites in the arid region." Bear Lake was one of the sites surveyed, and on July 29, 1889, under the authority of this legislation, "the public lands surrounding Bear Lake were segregated for use as a reservoir. This segregation was confirmed by Secretarial Order on August 18, 1894."[3] Further federal activity concerning Bear Lake occurred in 1902 when the Reclamation Service was established and reservoir administration functions were transferred to that organization. One of the first actions taken by this new organization was to direct a survey in 1902, conducted by W. G. Swendsen, a U.S. Reclamation Service surveyor, contemplating the use of Bear Lake as a water storage area for purposes of irrigation.[4] In 1903 "a temporary reclamation withdrawal was entered on the lands surrounding Bear Lake."[5]

Lucien L. Nunn was primarily responsible for hydroelectric developments on the Bear Lake—Bear River complex. He came to the Great Basin area with impressive credentials. He had built, during the winter of 1890, the Ames power plant at the confluence of Howard's Fork and Lake Fork on the San Miguel River, near Telluride, in southwestern Colorado. This plant, using the conventional water wheel and generator, provided alternating current at 3,000 volts. The electricity was passed over 2.6 miles of transmission lines to a 100 horsepower motor at the end of the line. This was "the first commercial high voltage, A C power transmission plant in the world."[6] This power plant, ultimately reaching a capacity of 1,740 horsepower, was acquired by the San Miguel Consolidated Gold Mining Company in 1892. Nunn, seeking larger water power sites and larger markets, expanded his interest into Utah, where in the mid-1890s he acquired power sites on the Provo and Logan rivers.[7] To combine his power interests in Colorado and Utah, Nunn incorporated the Telluride Power Company on February 19, 1900. At this stage the power plants were small, isolated facilities serving local mine, mill, or domestic requirements for electricity.[8]

Two developments related to hydroelectric power occurred about this same time that would have direct implications for developments on the Bear River. In 1895 new transmission methods for electrical energy were devised when it was discovered that if the voltage were increased from the customary 3,000 volts to 10,000 volts, the cost of transmission was reduced by over 90 percent.[9] The other development was a new plant at Ilium, Colorado, six miles below the Ames plant on the same stream. This plant, with a capacity of 1,610 horsepower, was completed in January 1902. It used waters from the tailrace of the Ames plant with a head of 501 feet.[10]

Nunn's interest in Bear Lake and Bear River came at the turn of the century. His own words clearly outline his plans for development of these two natural features.

> I had become impressed with the urgent need for reliable and con-
> tinuous water supply for the development of power in the territory
> served by the Telluride Power Company, and was continually investigat-
> ing the water sources of Utah and Idaho. In this way, [I] became famil-
> iar with the power possibilities of Bear Lake in connection with the fall
> near Grace, and at other places on the Bear River. I noticed, however,
> that, during the late irrigation season, Bear River was substantially dry
> at or near Grace. I was also impressed with the great benefit to irriga-
> tion in the Bear River Valley which would follow the development of
> Bear Lake storage. . . . [My interest was to appropriate] all of the unap-
> propriated waters of Bear River, to be stored in Bear Lake, and released
> for power, irrigation, and other beneficial purposes. . . . [The stored
> waters would be released] during the period of low water flow. . . . It
> was not possible, with the knowledge then at hand, to state definitely
> where all of the power would be developed or irrigation supplied; but
> Grace had been determined upon as a power site. . . . [In addition to
> the power plant, the work visualized included] . . . a reservoir
> project . . . namely, a large inlet canal, of capacity sufficient to im-
> pound all flood waters of Bear River, with releasing works to discharge
> these waters as they would be required for power and irrigation pur-
> poses. . . . The project was a large and comprehensive plan to take care
> of the constantly increasing demands for power and irrigation in the
> large mining district and the valleys above Salt Lake.[11]

P. N. Nunn, chief engineer of Telluride Power and brother of Lucien, was more conservative. In speaking of the possibilities in 1902 of a Bear Lake reservoir development, he stated: "It was known that something of a development was feasible, but the extent of that development was problematical."[12]

To resolve any doubts, one of the first requirements was to obtain additional information for the officials of Telluride Power about the Bear Lake and Bear River areas. J. C. Wheelon, perhaps sponsored independently, visited these areas in 1901–02 to investigate the feasibility of a project contemplating the storage of water in Bear Lake and the diversion of some of that water into Bear River for the purposes of both power and irrigation.[13] Furthering the quest for decision, Telluride Power conducted another survey of Bear Lake early in 1902.[14]

Because accomplishment of Nunn's plans was largely dependent on the acquisition of Bear Lake for use as a reservoir, problems relating to that subject were the first addressed. Included were such matters as appropriation of water, acquisition of privately owned property needed for the reservoir project, acquisition of right-of-way for establishment of a reservoir in instances of public domain, construction of a levee across the north shore of the lake to hold impounded waters, and excavation of both an inlet conduit from Bear River to the lake and an outlet conduit for discharge of stored waters back to the river.

Notices of appropriation of water, applicable to the reservoir concept, were posted on March 24 for 2,000 c.f.s. and on April 12, 1902, for 3,000 c.f.s. of Bear River water.[15] Appropriated was "the flood, waste, surplus, and unappropriated water of Bear River to be diverted from said river during the season of flood water and at all other times when diversion of said water will not interfere with prior appropriations or vested rights."

The appropriators under both notices were Lucien L. Nunn (general manager), William Story (counsel), Albert L. Woodhouse, William B. Searle (engineer), and Eldon P. Bacon, all officials of Telluride Power. Full compliance with the laws of the United States and of the states of Idaho and Utah was asserted. As required by law, the notices of appropriation were posted at the point of diversion from Bear River. Although the point of diversion was in Idaho, identical notices were placed at the post offices in both Laketown and Garden City, Utah, these "offices being the nearest within the State of Utah to said point of diversion." The two notices of March 24 and April 12 were notarized on April 11 and 22, respectively.

The purposes of these two appropriations were extremely broad. Included were

> preserving, saving and storing the . . . [appropriated water] in Bear
> Lake and thereafter reclaiming and withdrawing said stored water from
> said lake during the season of low water and utilizing the same down
> and along the valleys of the Bear River from the outlet of said lake to

the mouth of said river by supplying for such reasonable compensation as may be agreed upon at the various places in Bear Lake, Bannock and Oneida Counties in the State of Idaho and in Cache and Box Elder Counties in the State of Utah . . . such additional water as may be needed during said season of low water for power, irrigation, and all beneficial uses.

The means of diversion planned were described as "dams, head-gates, canals, flumes and such other works as may be found necessary or expedient in developing and completing this appropriation." The inlet canal, to convey water from the point of diversion to the lake for storage, was to be the principal construction facility and was to follow "the contour of the land so as to maintain a proper water grade, a distance of ten miles to the northerly end of Bear Lake." The inlet canal was planned to have a width of 150 feet and a depth of six feet. The outlet means for water stored in the lake was initially shown to proceed "northerly along the natural watercourse flowing from Bear Lake into Bear River, a distance of Sixteen miles." Results expected from development of the Bear Lake Reservoir included a storage capacity for more than an additional eighteen billion cubic feet of water, enough to irrigate or partially irrigate about 775,000 acres.

The points of diversion were different for the two water appropriations. This occurred because of a determination that "it would be necessary to divert the water from the river at a lower level in order to obviate the excess fall between the river and the lake."[16]

The notices of appropriation of water offered several features of special significance to all contemporary or future users of Bear River water. First, the officials of the power company were intent on acquisition of water rights that would permit the supplying of waters for a "reasonable compensation" to potential users over the entire course of the Bear River from Bear Lake to Great Salt Lake. The availability of "rental waters" was not merely incident to power developments. Second, the large quantity of water appropriated, totaling 5,000 c.f.s., was virtually preemptive of most of the remaining unappropriated waters of the Bear River. Third, there is a question about the need for two notices of appropriation, separated in time by only nineteen days. Was the second notice intended as an additional appropriation, or did it duplicate and enlarge the first—to correct deficiencies perceived in the initial appropriation action? William Story, attorney for Telluride Power, appeared to sup-

port, at least partially, the version of the second notice as being to
correct deficiencies in the first. He recalled that the second notice was
prepared, following some additional surveys by the power company
engineer, "for the purpose of fixing the point with greater definiteness,
or for the purpose of correcting the description, perhaps, of the point
of diversion."[17] However, a second "correction," one of much more
significance, was made relating to description of the outlet conduit.
On March 24 this conduit was described as proceeding "northerly
along the natural watercourse flowing from Bear Lake into Bear River,
a distance of Sixteen miles." After a lapse of only nineteen days, in the
notice of April 12 the description of the outlet conduit included the
modification "as far as practicable" with respect to following the natural watercourse. The position of the power company, strongly asserted
fifteen or so years later, denied the existence of a natural watercourse
from Bear Lake back to Bear River through which any significant
quantity of water could regularly flow. P. N. Nunn concluded, based
on readings from gauges installed by the power company, that "indica-
tions were that outside of the spring-freshet period there was no material discharge [from Bear Lake] at any time."[18] Whether there was
actually a natural outlet from Bear Lake would later have important
legal implications.

Other actions occurred coincident with the posting of the notice of
appropriation of 5,000 c.f.s. of Bear River water. The power company
began to acquire the land for construction of the inlet conduit, a canal
that came to be known as the Dingle Canal. In April and May 1902
virtually all the land needed for the canal right-of-way was purchased by
Telluride Power from the various local owners. A strip two hundred to
three hundred feet or more wide and from four to five miles in length
was thus obtained. Excavation commenced immediately and work on
this canal and the other features associated with the reservoir continued
steadily for the next ten years.[19]

Although purchase of land from local owners was accomplished with
little difficulty or delay, legal acquisition from the federal government of
the right-of-way for public domain and reservoir rights occupied Telluride Power officials for five years. Between 1889 and 1902 (confirmed in
1903), the public lands surrounding Bear Lake were set aside on the
basis of a "temporary reclamation withdrawal" by elements of the
United States Department of Interior. This precluded application of the

Right-of-Way Act of March 3, 1891, pertaining to rights-of-way for reservoirs.

Nunn and his associates continued—at considerable risk to their investments—their private development of Bear Lake in opposition to reclamation interest of the federal government. Based on their survey of Bear Lake in early 1902, maps and the right-of-way application for inlet and outlet for the reservoir were filed with the U.S. Land Offices at both Blackfoot, Idaho, and Salt Lake City. Because of the reclamation withdrawal, no action could be taken on the application. The power company, however, did not cease their effort, "but continued to present all the facts, as far as we could obtain them."[20] In this connection, L. L. Nunn wrote to Secretary of Interior Garfield in early 1907 stating the possibility of the development of sixty thousand electrical horsepower and irrigation of forty thousand acres through Nunn's project.[21]

Eventually, it became clear that the "Bureau of Reclamation [had] decided, finally, to put their initial efforts into other areas."[22] Telluride Power was able to arrange a hearing on April 1, 1907,[23] on their application for right-of-way and a reservoir at Bear Lake. Among those present were Secretary of Interior Garfield, Director Walcott of the Geological Survey, Director Newell of the Reclamation Service, Lucien Nunn, and Mr. Steigmeyer and Mr. Story, attorneys for Telluride Power. After a discussion stressing beneficial uses of the water both for power and irrigation, Secretary Garfield, without setting aside the earlier order withdrawing the Bear Lake Reservoir site from the public domain, nonetheless approved "the application made by Mr. Nunn for a right of way over Bear Lake north of Mud Lakes."[24] Conditions affecting or limiting the approval were not clear. Story recalled "quite a long discussion" of "places at which beneficial uses of the water could be made . . . both for power and for irrigation."[25] The Idaho Water Resource Board spoke of "the understanding that the right-of-way would be used 'for the development of power, as subsidiary to the main purpose of irrigation and drainage.' "[26] The right-of-way maps, prepared by Lucien Nunn and endorsed with Secretary Garfield's approval, were then filed by the power company at both the Blackfoot and Salt Lake City land offices.[27]

Even before the right-of-way problem concerning the federal lands surrounding Bear Lake had been resolved, Telluride Power commenced its first power development on the Bear River. During 1906 and 1907

preparations were under way at Grace, Idaho, just south of the Last
Chance Dam. The development plans included a dam to take advantage
of a narrow canyon and a natural falls location a short distance north of
Grace. The power plant, with its turbines and generators, was constructed in Black Canyon, five miles downstream from the dam. However, the water from the dam was carried overland four miles by penstock, in a straight line and with a fall of over five hundred feet, to the
plant. The power generated was to be transmitted at high voltage from
transformers at the Grace plant, via Logan and Salt Lake City, to the
mining districts of Bingham and Eureka. The preparations to implement
these plans included road construction from the village to both the dam
and the plant locations, as well as acquiring and storing supplies and
equipment, including seven million pounds of steel pipe.

The dam was constructed of rock-filled timber cribs, the same tech-
nique used (though on a much smaller scale) by Last Chance a few years
earlier. When completed, the dam was 40 feet high, 185 feet wide at
the bottom, and 340 feet wide along the crest. The upper fourth of the
penstock was of wooden stave construction, 8½ feet in diameter, the
remaining portion being made of riveted steel plates. At the plant

Utah Power and Light Company Dam
at Grace, Idaho (1982)

location at the terminus of the penstock were two 8,500 horsepower Allis-Chalmers turbines and two Westinghouse 2,300 volt, 5,500 kilowatt generators. Six transformers increased the voltage to 44,000 for economy in transmission.[28]

Appropriations of water for power purposes were entirely separate from the filings made in 1902 in connection with the Bear Lake Reservoir. On December 28, 1905, Lucien Nunn made application for a permit to use certain water of the Bear River for power purposes at the Grace installation. Dating from July 6, 1908, presumably the date that the Grace plant was placed into service, a "perpetual right" for the use of 500 c.f.s. was granted by the state of Idaho.[29]

The Grace plant, completed in 1908 at a cost of $7,176,949,[30] became the largest power-generating activity of the Telluride Power system. Advances in technology since construction of the Ames and Ilium plants in Colorado less than twenty years before were dramatic.

The work at Bear Lake Reservoir continued in the meantime.[31] Until late 1909 men and teams of horses performed the excavation work on the Dingle inlet channel, and by that date the Dingle Canal portion of the project was nearly completed from the intake on the Bear River to the junction of the canal with Mud Lake. In November 1909 a new excavation method—a large dipper-dredge—was in place and ready for operation. During 1910, 4,500 linear feet of additional excavation was done with the dredge on the Dingle Canal, 13,832 feet on the outlet canal, and 7,600 feet of levee along the north shore of Mud Lake. On May 3, 1911, installation of the headgates between the river and the Dingle Canal was begun and water was diverted into the canal for the first time three weeks later. The maximum quantity was measured as 402 second feet, the average flow being about 250 second feet. Diversion was stopped on July 13, 1911, by which time about 25,000 acre-feet of Bear River water had been stored in Mud and Bear lakes. The outlet channel was also used in 1911. On July 22 a temporary gate structure in the levee or dike across Mud Lake was opened and remained open until mid-October. "The total discharge . . . was about 41,000 acre-feet, resulting in a mean discharge of 250 second feet." The lake elevation decreased from 5,923.4 feet to 5,921 feet as a result of this discharge. The power company commenced further discharge on December 5 to compensate for a decline in river flow and the resulting shortage of water at the Grace plant.

Further improvements were made after 1911, including construction on the Bear River of a diverting dam made of rock in early 1912. The Rainbow Canal was constructed in 1914, having a capacity more than 4,000 c.f.s. It supplemented the Dingle Canal, which, although designed for a capacity of 2,000 c.f.s., was limited to 1,000 c.f.s. because of the steepness of the grade.[32] A river flow in excess of 4,000 c.f.s. was rare; thus, virtually the entire river was regularly diverted into Mud Lake.[33] By 1920 the use of the Dingle Canal was limited to "seasons of flood water."[34]

The reservoir facilities, when installations were completed about 1917, included two inlet channels, the Dingle Canal and the Rainbow Canal, to convey water from Bear River to Mud Lake. Diversion into Rainbow Canal was facilitated by the Stewart Dam, built in 1916. A dike existed between Mud Lake and Bear Lake. Bear Lake was filled by passing water through the sluiceway of Lifton Pumping Plant (which was completed in 1917) or through nearby headgates. Water was then released from Bear Lake back into Mud Lake by gravity through a sluiceway for water levels of elevations 5,923.65 feet or slightly higher to 5,920.65 feet or by pumping from the Lifton Pumping Plant for water levels of elevations 5,920.65 feet to 5,902.00 feet. Water released or pumped from Bear Lake flowed into Mud Lake where it was controlled by headgates at the Paris Dike and thence into the Bear Lake Outlet Canal. The released water again reached the Bear River channel about five miles below Stewart Dam, the point at which primary diversion occurred.[35]

The year 1912 brought several changes affecting the power company. A new organization, the Utah Power and Light Company (UP&L), was established, still under the direction of Lucien L. Nunn.[36] On May 8 that organization replaced Telluride Power as the operator of the Bear Lake Reservoir.[37] Incorporation of the new organization occurred on September 6,[38] and actions were quickly initiated to acquire the rights, resources, and properties of Telluride Power in Idaho and Utah.

Another significant event for 1912 was that UP&L applied for a third appropriation of water for storage in the Bear Lake Reservoir. On September 12, 3,000 c.f.s. was appropriated by application of the permit method. Sources of these appropriated waters were diversion from Bear River (2,500 c.f.s.) and the natural drainage into Bear Lake and Mud Lake (500 c.f.s.).[39]

The relationship between the newly incorporated UP&L and the Utah-Idaho Sugar Company (U&I) received close and early attention.

From the initiation of Nunn's plans for development of the Bear Lake—Bear River complex, Nunn considered the sugar company to be "a rival claimant of the Bear Lake Reservoir project."[40] Activities of U&I included some work in 1902 on an outlet channel from Bear Lake to Bear River.[41] Other activities of the sugar company on the Bear River also brought a further element of competition between the two corporations. The sugar company had, before late 1912, built its own dam in the Bear River canyon near Wheelon, Utah. This dam provided diversion of Bear River water for irrigation purposes and also permitted an associated hydroelectric development. This latter development consisted of the power generation plant, almost forty-five miles of transmission lines from the Wheelon location to a terminal station near Ogden, Utah, and twenty miles of local lines to various communities in Cache, Box Elder, and Weber counties in Utah.[42]

One of Nunn's actions oriented toward consolidation of the Bear Lake—Bear River development under his control was to attempt the elimination of U&I from involvement in both reservoir and power activities. How the sugar company was eliminated from the role of "rival claimant" on the Bear Lake Reservoir development is not clear. However, Nunn's plans for termination of U&I involvement on the Bear River were to be based on the expectation "to supply irrigation water to the sugar company tract in Box Alder {*sic*} County, Utah."[43] UP&L and U&I completed a contract to achieve these objectives on December 30, 1912. Under the terms of this contract the power company agreed to provide each year, commencing in 1913, 900 c.f.s. of water to the sugar company during the period May 1 to October 31, and 150 c.f.s. of water from November 1 to April 30. These quantities equated to a total of 328,468 acre-feet and 53,926 acre-feet for the periods May 1 to October 31, and November 1 to April 30, respectively.[44] The water was to be delivered at the canal headgates belonging to the sugar company. Maintenance, operation, and repair of the dam, diversion works, and outlets to the sugar company headgates were responsibilities of the power company. The sugar company was authorized to generate power for pumping, if incidentally generated in the fall of the canals. The power company was also responsible to provide electricity for sugar factories and other major sugar company requirements "at as low a price as the same shall be sold and delivered to any other customers for similar service in the State of Utah."[45]

The sugar company surrendered considerable apparent real value in

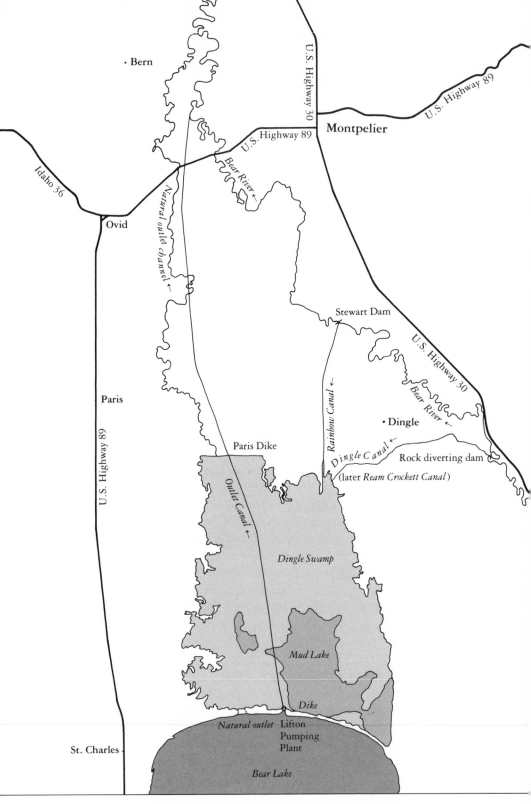

Bear Lake Reservoir and Facilities

exchange for the guaranteed availability of water for irrigation and the other considerations. These included (1) the Wheelon main dam; (2) the hydroelectric plant, to include a new generator not yet installed, and all associated lines and facilities; (3) 640 acres of land associated with the dam, diversion works, and canals; (4) rights and easements to flood certain lands upstream from the dam; (5) "all its right, title, interest and estate . . . in and to the waters and the use of waters of the Bear River,"[46] except that, which under the contract, the power company was responsible to supply; and (6) a payment annually of $4,000 toward the maintenance of the diversion works.

The element of competition between the power company and the sugar company was ended by Section X of the contract.

> All controversies between the parties hereto . . . respecting the right to the use of the waters of the Bear Lake and Bear River, are hereby settled, and any and all claims of either against the other up to this date, are hereby released.[47]

Thus was completed what the Idaho Water Resource Board described, in connection with the concept of an irrigation reserve to be maintained in Bear Lake, as the only instance of "a firm, continuing contract . . . for use of Bear Lake water."[48] The authority for UP&L to make such a commitment rested on an unchallenged assumption of their "exclusive rights to Bear Lake." The results of this contract were important. First, total control over hydroelectric developments on the Bear River was conferred on UP&L. Second, that company was taken "by the back door" into the business of providing electricity to the communities in the vicinity of the generating plants when the power company acquired the Wheelon plant and the associated local lines serving the communities of Cache Junction, Fielding, Riverside, Garland, and Tremonton. Even at Grace, Idaho, when their plant was completed in 1908, UP&L did not market electricity to consumers in that village and vicinity. A distribution system for Grace was constructed in 1909 by three local residents who bought power wholesale from the power company, reduced the voltage with transformers, and sold it to local consumers. Not until 1929 was this local distribution method for sale of electricity purchased by UP&L.[49]

UP&L had a broad vision for development of the Bear Lake—Bear River complex. The vital importance of the Bear Lake Reservoir to their plans continued to be well recognized. However, major conceptual changes were made in the area of hydroelectric developments. Under this new concept, all hydroelectric plants, present and future, on the Bear River were to be interconnected by new, long transmission lines and specially constructed terminal stations and load centers.[50] UP&L visualized an integrated system rather than the method of isolated, independently functioning plants previously followed.

It is appropriate to consider briefly the growth of the UP&L "system" during the twenty years following completion of the first hydroelectric plant on the Bear River at Grace, Idaho. Table 2 summarizes the developments.[51] (See also map on page 7.)

The power company made improvements in the system from time to time that promoted greater efficiency or increased generating capacity. For example, in 1913 and 1914 major improvements were made at the Grace plant.[52] An additional eleven-foot diameter penstock was built from wooden staves. New turbines and generators increased the plant capacity to 33,000 kilowatts. New transmission lines, mounted on steel towers, covered the 135 miles from the Grace plant to a terminal just west of Salt Lake City, which was established to accept power from all the Bear River generating stations. A further modification at the Grace plant in 1923 increased its capacity an additional 11,000 kilowatts. The Oneida plant—started before the passing of Telluride Power in 1912—was also modified. Both in 1916 and 1920 a generating capability of 10,000 kilowatts was added.

The Cove plant was an interesting development, reminiscent of the Ilium, Colorado plant in 1902. In both cases water was taken from the tailraces of existing plants to operate new generating facilities. From the Grace plant, water "leaving the tailrace . . . tumbled down a river bed which dropped almost a hundred feet in the next mile and a half. . . . By taking the water directly from the tailrace, and running it parallel to the river in a huge wooden flume, another 7,500 kw could be generated before returning the water to the river."[53]

The Cutler plant, which came into service in 1927, was built to replace the Wheelon plant, acquired in 1912 from U&I.

There were, of course, appropriate and timely filings for Bear River water to operate the several hydroelectric plants. As finally decreed, waters of the Bear River appropriated for power purposes included an additional 500 c.f.s. for the Grace plant, thus giving a total power right for that location of 1,000 c.f.s. Other Bear River power rights, resulting from appropriations before 1920, were distributed as follows: Oneida, 2,500 c.f.s.; Cove, 1,500 c.f.s.; and Wheelon, 1,040 c.f.s.[54] The use of

UP&L Power Facilities

Year	Plant	Initial Capacity (kw)	Modified Capacity (kw)
1908	Grace	11,000	44,000
1910	Paris Creek	650	650
1912	Riverdale	3,750	3,750
1915	Oneida	10,000	30,000
1917	Cove	7,500	7,500
1924	Soda (Alexander)	14,000	14,000
1927	Cutler (Wheelon)	30,000	30,000

waters for power generation was less significant to river flow than use for irrigation, since the water was returned to the river channel after leaving the power plant turbines.

The completion of hydroelectric developments on the Bear River also had the effect of creating a system of reservoirs, reservoirs that, in some cases, served uses other than those associated solely with the generation of power. Supplementary flood control and diversion of water from the river for irrigation were examples of such uses. Reservoir capacities were impressive. Bear Lake Reservoir was considered to have a capacity of 1,450,000 acre-feet; Oneida, 11,500 acre-feet; Soda, 11,800 acre-feet; and Cutler, 12,700 acre-feet.[55] In addition, the Grace dam created a small but useful reservoir.

The power developments on the Bear River and Bear Lake were major accomplishments, applying a still largely experimental technology to a major Great Basin hydrological complex. In a nineteen-year period (1908 to 1927) UP&L and its predecessor organization, Telluride Power, built, and in some cases modified, seven hydroelectric plants possessing a total generating capacity of almost 130,000 kilowatts. The roles of both Lucien L. Nunn and his first Bear River plant at Grace, Idaho, were important to the further economic development of the area; both the man and his work deserve the attention of history. But, the principal achievement was the establishment of an integrated system, involving both the generation of electric power and its efficient distribution over considerable distances to high-demand market areas.

CHAPTER SEVEN

The Bear River
Water Case

The Case Defined

For three years (mid-1917 to mid-1920), a court case that came to be known as the Bear River Water Case was in process in the District Court of the United States for the District of Idaho, Eastern Division. The case was designated Equity No. 203, Utah Power and Light Company, a corporation, plaintiff, versus the Last Chance Canal Company, Limited, et al.

There were a total of 531 defendants in the action, including the Last Chance Canal Company and its seven subsidiary canal companies. Of the remaining defendants, about 12 percent were other corporations, companies, partnerships, financial institutions, and churches. Nearly half of these organizations were associated with the common use of Bear River system water for irrigation. Almost 87 percent of the defendants were individuals, many with minor claims to water.

The Last Chance Canal Company and its subsidiaries were identified by UP&L as the "principal adversaries in this litigation" by virtue of the "geographical position, above all power plants, and irrigation diversions in which the plaintiff has any interest, and of the fact that it is, next to the Utah-Idaho Sugar Company, the largest system upon the river."[1] Accepting this logic, all further discussion emphasizes the relationship of the power company as plaintiff to the Last Chance Canal Company and its subsidiaries as defendants. This is done to avoid unnecessary repetition and to further the purpose of this writing as a case study of irrigation versus power and commercial interests on the Bear River.

The motivations causing the power company to file suit have been suggested by internal power company developments and by existing law. Major expansion of the hydroelectric operations on the Bear River, commencing in 1912 with the organization of UP&L and the signing of the broad contract with U&I, brought a need for a clear definition of water right, both on the river and at Bear Lake. A clear bestowal of these rights could come only from adjudication, as prescribed by Idaho water law.[2]

From the judicial processes of adjudication would come the judge's decree defining the priorities and extents of the adjudicated water rights. The elements of the contending claims considered by the court included priority of appropriation, extent of the right in quantity, and the extent of the right in time (see chapter three). Filing dates and specific language of the several apppropriations were important factors in determining priority. Factors influencing the evaluation of the extent of the right in quantity included the capacities of the diversion works as actually constructed and economy and reasonableness of use of water—the so-called "duty of water" concept. The extent of the right in time was defined by statute in the instance of beneficial use for irrigation as a seasonable applicability beginning and ending on specific dates.

Last Chance officials were notified in June 1917 of the filing by UP&L of the suit "to have the waters of the Bear River decreed according to their rights."[3] The canal company decided at that time to seek the expert assistance of an attorney and a "chief engineer." W. G. Swendsen, who had conducted a survey of Bear Lake in 1902 as an employee of the U.S. Reclamation Service and who later by June 1919 held office as commissioner of reclamation for the state of Idaho,[4] was hired for the chief engineer position. His principal tasks were to measure the water and plat the land served by the Last Chance system. J. H. Peterson, of the Pocatello firm of Peterson and Coffin, became the attorney.[5] This assistance was expensive. Swendsen received $8,036 and Peterson received $16,450 for services during the case.[6]

The Judge[7]

On the bench for this case was Judge Frank S. Dietrich—a vital force as the individual who would interpret and apply the law. Dietrich was born in 1863 and was raised on a farm in Kansas. He completed one year at Ottawa University in his native state before transferring to Brown University in Rhode Island. At Brown he earned A.B. and M.A. degrees (in 1887 and 1890, respectively), interrupting his studies there to teach history, political economy, and Latin at Ottawa University. He also began to study law. By 1891 he had moved to Idaho where he continued "reading law." In January 1892, Dietrich, at the age of twenty-nine, was admitted to the Idaho bar.

In 1907 he was appointed as Judge, District Court of the United States for the District of Idaho. Thus, when the Bear River Water Case was filed before his court, he was a mature jurist who had been in the legal profession for twenty-five years and had been a judge for ten of those years. Judge Dietrich was highly respected in Idaho. Integrity, a sense of fairness, and a veneration for the law characterized his actions. Recognition of his judicial quality continued throughout his life. He is reported to have declined appointment as President Harding's attorney general so that he could remain on the bench. In 1926 he was appointed as Judge, United States Circuit Court of Appeals for the Ninth Circuit. Until his death in 1930 he aspired to appointment to the United States Supreme Court.

Legal Skirmishing

The initial hearing on the motion, scheduled by the court for July 10, 1917, was delayed at the request of the defendants' attorney,[8] and a variety of other delays contributed toward a lengthy court action. The chief engineer for Last Chance requested delay in submitting proof "as to their original appropriation, diversion, and application of water." Among his reasons was the nonavailability of skilled engineers owing to World War I.[9] Further delay in the proceedings was occasioned in late 1918 because of Judge Dietrich's illness with typhoid fever.[10]

On June 29, 1918, UP&L requested the appointment of a commissioner or special master "to protect the plaintiff in the use and enjoyment of the water discharged from its Bear Lake Reservoir in excess of the quantity being diverted into said reservoir from Bear River." Judge Dietrich met this request on July 12.[11] Continuation of the commissioner position was especially important in 1919, a year when UP&L anticipated that much water would have to be pumped from Bear Lake Reservoir because the flow of the Bear River was well below normal.[12]

The Arguments: Last Chance Canal Company

After preliminaries that lasted two years, the case, Equity 203, *Utah Power and Light Company v. The Last Chance Canal Company, et al.*, was scheduled to be heard in District Court of the United States, District of Idaho, Eastern Division, at Pocatello on June 23, 1919.[13] The position of Last Chance, as a principal defendant, can be reconstructed with some confidence from a summary of historical circumstances and from particulars advanced for emphasis in certain of the legal documentation.

The first requirement for Last Chance, to establish its legal position, was to identify positively its appropriations of the public waters of the state of Idaho flowing in the Bear River. As discussed previously, appropriation—demonstrated by the diversion of water from a natural watercourse—consisted of two aspects: the posting of the notices of appropriation (or later, the obtaining of the necessary permits) and the construction of the means of diversion. One method used by the Last Chance attorney for gaining association with appropriations was through "predecessors in interest." Under this concept, Last Chance attempted to show that the company had its origins in previous efforts to irrigate the same lands with Bear River water diverted from the same or near the same locations along the river. From predecessors in interest—that is, from the "pre-Last Chance" earliest irrigation efforts in Gem Valley— claim was made for appropriations of a total of 6,616 c.f.s. of water, with priority as early as November 1885 for 100 c.f.s. of this total.[14] Whereas full compliance with the law was asserted in the posting of notices for the predecessors in interest appropriations, evidence of construction of the means for successful diversion was, in most cases, extremely weak. Only in the case of an appropriation of 86 c.f.s. on September 2, 1889, did clear evidence exist that water for irrigation had been diverted onto the land.[15] But within six years that effort had been abandoned.

Two appropriations were made directly by the Last Chance Canal Company or its officials. The first, for 400 c.f.s., occurred in early 1897. Recall that the exact date of that action was unclear as both February 14[16] and February 24[17] were possibilities. A second filing for 600 c.f.s. of Bear River water occurred in May 1901. Exact dates were again unclear, and for the same reason—as both May 11 and May 14 were recorded.[18]

Contents of the notices of appropriation other than dates and quantities of water were also important. The proposed application of the water to be diverted for irrigation and culinary purposes was an unchallenged beneficial use. Descriptions of the point of diversion and of the places of intended use, however, were problems. The descriptions were vague and imprecise, particularly in the 1897 filing. A knowledgeable witness for the Last Chance Canal Company, John Trappett, stated an intent to irrigate forty thousand acres with the water appropriated in 1897.[19] With the 1901 filing, however, the estimate of acreage planned to be irrigated had increased to seventy-five thousand acres.[20] In addition, descriptions of the planned means of diversion were disturbingly broad. The 1897 application described the means of diversion as a dam, a flume three thousand feet in length, and two canals, each eight miles long. By 1901 the description included a suggestion of a system of ditches and branches having an aggregate length of one hundred miles. Both filings included a time limit for completion within five years, which limitation was either required by law or specifically accepted by the filing commitment.

The 1897 appropriation required one further action. This filing had been made by three individuals representing, erroneously, the Last Chance Irrigation Company. A quit claim deed was prepared on October 22, 1901, to pass those interests to the Last Chance Canal Company.[21]

Last Chance commenced excavation and construction of the means of diversion within sixty days of the appropriation of water, as required. The work proceeded "diligently and uninterruptedly," except when temporarily interrupted by adverse weather.[22] Two days before the five-year deadline for completion of the diversion in 1902, water had been placed, it was hoped, "onto the land far enough to comply with the . . . law."[23] At this stage water diversion had been accomplished by construction of a diversion dam, headgates, a flume 275 feet long, 1,100 feet of open ditch, another flume 1,800 feet long, 1,800 feet of open ditch, and the headgates leading to the East Branch Canal, one of two on the east side of the river.[24] Construction of an additional flume in March and April 1902 conveyed water to the west side of Bear River where the Central, North Extension, and Tanner canals were supplied.[25]

To summarize, by 1902 Last Chance had diverted water from the Bear River by construction of approximately one and one-fourth miles of works, having a capacity of 450 c.f.s.[26] Some irrigation was practiced that year,[27] but the extent is not known. Thirty-five hundred acres were estimated as under irrigation by 1903.[28]

Recognizing, perhaps, that a major deviation existed between what had actually been built and what had been described in the two notices of appropriation (1897 and 1901), some explanation seemed to be in order. The Last Chance attorney advanced a concept of the Last Chance "system," which was described as "one single, inter-dependent system of canals," a cooperative organization for the purpose of furnishing water for irrigation, domestic, and other purposes.[29] The implication was obvious; once the Last Chance Canal had conveyed water to points of delivery to branch canals, built for convenience in operation and maintenance and organized as subsidiary or secondary corporations, the legal obligation of the Last Chance Canal Company had been fulfilled.

Three additional water appropriations—after Last Chance had been in operation for nearly a decade—made by the Bench Canal Company and the Tanner Canal Company closed the appropriation "chapter" in Last Chance history. On August 9 and December 31, 1909, the Bench Canal Company appropriated 138.16 c.f.s. and 25.6 c.f.s., respectively.[30] On July 29, 1910, the Tanner Canal Company appropriated 54.0 c.f.s.[31] In both cases the circumstances of the diversion were similar. Points of diversion different from that of the Last Chance Canal were selected. The two companies took advantage of the reservoir behind UP&L's dam just north of Grace rather than construct their own diversion dams. The appropriated waters were diverted from the east side of the reservoir into a new canal, the Bench "B", and from the west side into the Tanner "B" Canal. In both instances the "B" canals were short in length and emptied their waters into the main Bench and Tanner canals. In May 1917, the Last Chance board of directors voted to "take over the works, water rights, and holdings" of the "B" canals. They paid for the canals with 4,500 shares of Last Chance stock, divided proportionally to the amounts of water appropriated.[32]

The portrayal of the circumstances behind the Bench "B" and Tanner "B" appropriations was perplexing. W. G. Swendsen, engineer for the Last Chance Canal Company, recommended that these new appropriations and the associated construction of means of diversion be considered as an extension of the Last Chance system under the concept of the "due diligence law." This interpretation would increase the likelihood of entitlement to the 1,000 c.f.s. already appropriated by Last Chance and would establish priority as of the original dates of appropriation.[33] Although the Swendsen position was officially adopted by Last Chance,[34] his view was not fully accepted. A potential witness on behalf of Last Chance regarded such a position as "embarrassing and false." Further, there was some feeling that the Bench "B" and Tanner "B" filings might be competitive and thus detrimental to Last Chance interests. Associated with this belief was a conviction that the Bench "B" and Tanner "B" interests should not be introduced into the Last Chance case. Still hoped for was a decree giving Last Chance "our 1000 feet (c.f.s.) as of original date."[35]

The Last Chance system had continued to change. The system capacity by 1905 had increased nearly 16 percent, up from 450 c.f.s. in 1902 to 520 c.f.s. This increase in capacity, however, was not sustained. In 1918 the capacity was 428.5 c.f.s.; the number of acres irrigated was estimated to have been about 29,000. By 1919 the system included ninety miles of main canals and 127.75 miles of laterals.[36]

The Last Chance was concerned with two points pertaining to the quantity of water. The first applied to economy and reasonableness in the use of water, the so-called "duty of water." Because of soil and climate considerations, it was "discovered that it took more water to water that land than a person would naturally think."[37] On the basis of this logic and the possibilities of further expansion of the Last Chance system, Last Chance officials concluded that all the waters claimed were "necessary to the successful cultivation and irrigation of the lands . . . described and other lands entitled thereto and susceptible of irrigation from said system of works." This argument is not clear. The 1,000 c.f.s. of water directly appropriated for or by Last Chance should have irrigated about 50,000 acres of land, applying the rough "rule-of-thumb" guide to duty of water of one cubic foot per second to irrigate fifty acres. Since irrigation of almost 49,000 acres was asserted,[38] apparently further claim to the 6,616 c.f.s. of water from predecessors in interest was abandoned. The second point was a bold statement bringing "due diligence" growth and expansion into consideration. Last Chance stated that their system had "annually and for many years past diverted an increasing quantity of the flow of . . . [Bear] river . . . [and] have put such water to a beneficial use."[39]

In addition to defending their legal position concerning Bear River water appropriation, Last Chance filed a cross complaint against UP&L for certain power company actions in establishing the reservoir at Bear Lake. Noting that the power company had constructed a dike along the north edge of Mud Lake between that lake and Bear River, and a deep canal leading from the river to Mud Lake, Last Chance charged that these works did, in fact, "hinder, retard and impede the natural outflow, run-off and drain from . . . Mud Lake and . . . Bear Lake into . . . Bear River."[40] According to the Last Chance view, the natural outlet from Bear Lake to Bear River—being small, crooked, and congested—would freeze in winter and thus impound large amounts of water naturally draining into the Mud and Bear lakes during early spring floods. This natural reservoir, holding and conserving the drainage water, would then slowly and gradually discharge the naturally stored water during the entire summer for flow down the Bear River. Irrigators planning to use the natural flow of the Bear River considered the reservoir facilities unnecessary and unwise.

Following the Last Chance legal arguments came certain requests to the court. First, UP&L should be granted "no relief . . . which impairs, conflicts with or prejudices the right of . . . [Last Chance] to the waters so claimed by it." Second, Last Chance claims to Bear River waters earlier appropriated should be decreed. And third, Last Chance should recover its costs incident to the court case.[41]

It was crucial that the court accept that Last Chance diversion efforts between 1897 and 1902 had been according to law and had totally succeeded. Under the Last Chance *system* concept, the added facilities and expansion of the subsidiary canals after 1902 in no way detracted from the fact that water had been diverted from the natural watercourse onto Gem Valley lands for the beneficial use of irrigation. Appropriation had already been accomplished. The task of the court, therefore, was to judge, to adjudicate, to decree, the elements of the right, as compared to those of others, to the water claimed or appropriated.

As to priority of appropriation, the Last Chance Canal Company was in an extremely good position. The doctrine of relation applied since the statutory method of appropriation had been followed. Under this doctrine, if the works were completed as required, the priority of the right could be associated with the date of the posting of the notice of appropriation. The filing of 1897 (specifically February 14, February 24, or March 1) for 400 c.f.s. of water was secondary in priority to only 436.1 c.f.s. appropriated by other claimants on the entire portion of the Bear River being considered by Dietrich's court. Stated differently, the Last Chance Canal Company's position as to priority was secondary only to 5.7 percent of all main river water appropriations eventually decreed.[42]

Consideration of the Last Chance claims as to extent in quantity was, as hinted in chapter three, more complex and much less certain. The Last Chance claims of 1897 and 1901 totalled 1,000 c.f.s., but the capacity of the diversion works constructed never approached those quantities. Reflected by percentages of system capacity to water claimed, this principal limiting factor to the Last Chance position was crucial. For example, in 1902 actual diversion was only 45 percent of the 1,000 c.f.s. claimed. Diversion had increased to 52 percent of the claimed amount in 1905, and then decreased to 42.9 percent in 1918. Last Chance was also vulnerable to criticism for using water in excess of that suggested by economy and reasonableness. Adherence to a view favorable to the "duty of water" concept could have led to a conclusion that all or a portion of this scarce and precious resource could be used elsewhere to greater public advantage than within the Last Chance system.

The final element to be considered in the adjudication of the water appropriations by the Last Chance Canal Company was extent in time. Since the beneficial uses specified for the appropriated waters were both for irrigation and for culinary or domestic purposes, the right in time for the irrigation portion was limited to the duration of the irrigation season as defined by law. The portion of appropriated waters for culinary or domestic purposes would not be so limited.

The Arguments: Utah Power and Light Company

Historical events with respect to Bear Lake and Bear River participated in by Telluride Power and its successor corporation in 1912, UP&L, would, as had been true for the Last Chance Canal Company, largely define the legal position of the power company—the plaintiff in the case. These events concerned two separate but closely related activities: establishment of a large, controlled reservoir at Bear Lake and construction of hydroelectric plants at selected locations on the Bear River.

The power company appropriated 2,000 c.f.s. of Bear River water for storage in Bear Lake on March 24, and another 3,000 c.f.s. on April 12, 1902, for the same purpose, actions that led to the establishment of Bear Lake Reservoir.[43] Land acquisition for the reservoir project started in early 1902 when the company purchased privately owned land needed for the project.[44] UP&L did not acquire the necessary right-of-way over federal lands until April 1, 1907.[45] However, the company started construction of the means of river diversion into the lake without waiting for approval from the federal government.[46] Between 1902 and 1911, dikes were constructed north of Mud Lake and between Mud Lake and Bear Lake to provide a means of controlling any water impounded. Additionally, an inlet canal, the Dingle Canal, was excavated to divert water from Bear River to the north end of Mud Lake. The capacity was limited because of grade problems and the inlet canal probably never reached a capability to carry over 50 percent of the water appropriated.[47] An outlet conduit, following roughly an existing natural outlet from Bear Lake to Bear River, was improved for water flow. Through temporary headgate structures, the first water was diverted into the Dingle Canal on May 24, 1911. This water was subsequently passed from Mud Lake and impounded in Bear Lake. The first release of stored water, again through temporary headgates, occurred on July 22, 1911.[48] As the reservoir project progressed, the power company made several improvements. In 1912 a rock dam was constructed in the river to aid in the diversion of water. A much improved inlet conduit, the Rainbow Canal, was completed in 1914 to supplement and finally virtually to replace the deficient Dingle Canal. The Stewart Dam was finished in 1916 to aid in diverting river waters into the Rainbow Canal. The final major improvement was the Lifton Pumping Plant, operational in 1917.[49] From this facility considerable stored water that could not leave the reservoir by gravity flow could be pumped into Mud Lake, from whence it could flow by gravity into the outlet channel for eventual return to the natural watercourse of the Bear River.

While reservoir construction was in progress, the power company completed its first hydroelectric plant on the Bear River at Grace, Idaho, in 1908.[50] Water for power purposes at Grace was appropriated December 28, 1905; 500 c.f.s. was granted by the state of Idaho in 1908.[51] Other hydroelectric developments along the Bear River and some of its tributaries quickly followed completion of the Grace plant. A plant was operational on Paris Creek in 1910, at Riverdale in 1912, at Oneida (near Preston, Idaho) in 1915, and at Cove (associated closely with the Grace plant) in 1917. Two other hydroelectric plants were built on the Bear River after the Bear River Water Case—Soda (Alexander) in 1924, and Cutler (Wheelon) in 1927.[52] Water for all these power plants, operational after 1908, was appropriated by the power companies concerned, but these appropriations did not become major issues at the time since the water used, after passing through each plant's turbines, was returned virtually undiminished to the natural watercourse of the Bear River. Provision of water in dependable supply, even during periods of natural low water flow, was, of course, of major concern to the operators of the Bear Lake Reservoir and the Lifton Pumping Plant.

On December 30, 1912, a contract between U&I and UP&L[53] gave the power company an extensive and permanent involvement both in

measures to insure a regular, predictable flow of water in the Bear River
and in its use for irrigation purposes. The contract was complex. U&I
had done some early work at Bear Lake,[54] having recognized the lake's
value to insure seasonal availability of water on the lower portions of the
Bear River. The sugar company had appropriated a total of 509 c.f.s. of
Bear River water in 1889, 1901, and 1914 to irrigate its beet fields in
Cache Valley in northern Utah.[55] U&I also had built a hydroelectric
plant at Wheelon, Utah, an excellent location on the Bear River. With
the power generated at the Wheelon plant, the sugar company provided
electricity to domestic consumers in various communities of northern
Utah.[56] In exchange for the Wheelon hydroelectric plant, the Bear River
water appropriations made by U&I, and any claims to reservoir rights on
Bear Lake, UP&L contracted to provide annually to U&I at their canals
near Wheelon 900 c.f.s. of water during the period May 1 to October
31 and 150 c.f.s. for the remainder of each year.[57]

The contractual commitment of UP&L to supply large quantities of
water to U&I, together with hydroelectric facilities from Grace to
Wheelon on the Bear River at the time of the court action, gave the
power company an interest covering much of the length of the river. A
further consideration dictating an "entire river" concern was the desire
to supply water, withdrawn from Bear Lake storage during the season of
low water, for "reasonable compensation"[58] to the maximum potential
market. For these reasons the legal strategy of UP&L, in addition to
substantiating their own claims by performance, became one of diminu-
tion of rival claims, especially large claims, on the upper portions of
Bear River. As pertained to Last Chance, UP&L challenged the predecessor in interest claims and all aspects of Last Chance's five appropriations
judged acceptable for substantial consideration—the priority, the quantity, and the period of beneficial use.

The claims stemming from predecessors in interest were disposed of
quickly in argument as "surely not seriously urged" and as lacking "any
proof of attempted compliance with law." These claims were described
as "simply evidences of abandoned projects by early settlers," with no
attempt being made "either to show when or how the Last Chance Canal
Company succeeded to any of these archaic notices of appropriation."[59]

Five claims were addressed for "substantial consideration."[60] These
were the Last Chance appropriations of 1897 and 1901, the two appropriations of the Bench Canal Company of 1909, and the Tanner Canal

Company appropriation of 1910. These appropriations represented a total of 1,217.76 c.f.s. of water from the Bear River, of which 1,000 c.f.s. were covered by the Last Chance appropriations.

The priorities of the two Last Chance appropriations were challenged. In this connection, the importance of the applicability of the doctrine of relation cannot be overemphasized. If this doctrine applied, a benefit of five years in priority of the appropriations would accrue to Last Chance. Not surprisingly, the reaction of UP&L was strongly negative. As expressed by their attorney, "to secure the benefit of the doctrine of relation . . . a greater measure of diligence, and some approximation to compliance with the statutes and established rule of law governing such appropriation, must be shown."[61] Without strict compliance with statute details, the doctrine of relation would not apply and the date of beneficial use of the water would determine the priority. Under these circumstances the Last Chance rights would be *after* the power company's claim of storage rights of March 24, 1902, as well as after various irrigator rights initiated between 1897 and 1902.

The attorney for the plaintiff argued that Last Chance's 1897 notice of water right lacked sufficiency to establish adequate claim to the water appropriated. Deficiencies were noted in descriptions of the point and means of diversion, and the places of intended use. Inconsistencies in the dates alleged for posting of the notice were mentioned, as were omissions in statements required by the statute.[62] The attorney further asserted that "the works were not constructed within the time or with the diligence which the statute demands."[63] The first point, an assertion of failure to complete within five years, brought a sharp disagreement with the concept of the Last Chance "system," as favored by the irrigators. The UP&L position was that the works, as visualized and described by the notice of water right, were not completed when a diversion dam and slightly over one mile of flume and ditch had been constructed. These works did not conduct waters to the places of intended use, an achievement requisite to completion of the appropriation. The works constructed by Last Chance were not as described. Delivery of water to "the place where head works of other canals, to be later constructed, will be, is not equivalent to the construction of such canals." The argument continued concerning the subsidiary canals: they were "never constructed according to any preconceived plan, to any given capacity which can be related to the notice, or to any proportionate part of the quantity of water appropriated." It was, in the opinion of UP&L, a subterfuge to argue that the terminus of a short canal was properly considered the place of intended use. The place of intended use could not be "the head works of the canals that were constructed in later years by the other seven corporate defendants, which claim to be a part of the Last Chance group." The convenient method of "simply changing . . . plans and constructing only a fractional part of the main canal and getting others to construct the remaining portion of the system after the expiration of the period allowed by the statute" was deplored. Under such circumstances the subsidiary corporations of the Last Chance system established "no title to the notice or rights under the notice, and the work of construction on the large canals and laterals owned by the separate corporations was not commenced until after the five years had expired." The power company considered that there was no "authority for holding that this vast irrigation system, held under eight separate and distinct ownerships and constructed by eight separate and distinct corporations over a period of twenty years, may not be . . . decreed to have been completed by the Last Chance Canal Company within five years after the posting of the notice in February, 1897."[64]

In addition to the allegation of failure to complete the diversion within the maximum allowable time, the power company also argued that the canal company had failed to carry out the work in a diligent and continuous manner. This argument was especially directed toward the 1901 filing, wherein a failure to start work within sixty days of the posting of the appropriation notice was stated, as well as a statement of total failure to enlarge the Last Chance Main Canal in consideration of enlargement of the total appropriation from 400 c.f.s. to 1,000 c.f.s.[65] Indeed, the power company suggested that the basis of the Last Chance 1901 filing was incident to a corporate reorganization and was a recognition of the Last Chance failure to meet the legal requirements of the 1897 filing.[66]

The priorities associated with the 1909 and 1910 water appropriations by the Bench and Tanner canal companies were not challenged by UP&L, provided that these filings were regarded as appropriations entirely by independent corporations and producing entirely independent rights. The power company was firm in the belief that these appropriations could not be related back to the 1897 or 1901 Last Chance filings or to "any kind of earlier diligence or actual use rights."[67]

The attack of UP&L on the appropriations of the Last Chance Canal Company was most effective in consideration of the extent in quantity.

Even though the two appropriations totalled 1,000 c.f.s., the capacity of the main canal never approached that value. In 1902 the capacity was 450 c.f.s. and in 1918 it was 428.5 c.f.s.,[68] thus giving an average capacity over the period of about 440 c.f.s. The early water law was explicit on the matter of capacity of the works constructed. The extent in quantity of an appropriation was limited to that which the "works are capable of conducting, and not exceeding the quantity claimed."[69]

Another factor was applied in evaluating the extent in quantity of an appropriation and this was a consideration of the needs of the appropriator. It was incumbent on any appropriator to exercise economy and reasonableness in the use of water. As previously discussed, this concept, known as "duty of water," was by 1903 defined as a limitation on diversion of water for irrigation purposes to an amount not to exceed one cubic foot per second for each fifty acres to be irrigated, unless needs in excess of these amounts could be demonstrated.[70] UP&L provided comparative data relating to duty of water on the Bear River. Large irrigators on the lower portions of the river were identified with a use factor of one cubic foot per second to irrigate eighty acres. An expert witness suggested a ratio of one cubic foot per second of water for sixty acres, but further suggested that the acres to be irrigated could be increased to eighty with improved management.[71] Based on what must have been an exasperating analysis to the Last Chance irrigators, complete with a lengthy and highly scientific treatise on plant biology and water requirements, the power company attorneys concluded that the normal duty of water ratio should be one cubic foot per second for *one hundred* acres, and never should the acreage to be irrigated with that quantity of water be less than eighty.[72]

The acreage under irrigation was the next value to the considered in the duty of water equation. A total of 21,449 acres was suggested by UP&L as under irrigation from Last Chance canals in 1917. Another analysis by the power company legal staff suggested a low total acreage intended to be irrigated. In this instance it was noted that the 45,000 shares of Last Chance stock were distributed on the basis of three shares per acre. Thus, the total acreage ever intended to be irrigated could not exceed 15,000.[73]

Using the acreage under irrigation value of 15,000 and the fact that the state recognized a duty of water ratio of one cubic foot per second for fifty acres, an argument was advanced for an extent in quantity of the Last Chance appropriations of 300 c.f.s. If the 1917 irrigated acreage total of 21,449 acres was accepted and the various possible duty-of-water ratios mentioned were applied, the extent in quantity could be calculated as some value from a low of about 214.5 c.f.s. to a high of almost 429 c.f.s. Or, to use the two values of the acreages under irrigation (21,449 and 15,000 acres, respectively) and the average capacity of the Last Chance main canal in 1918 (440 c.f.s.) in a different way, the Last Chance duty-of-water ratios could be computed as one cubic foot per second for 48.7 acres or one cubic foot per second for 34.1 acres. To the power company these use factors suggested a waste of water.

Such waste was attributed by the power company's attorneys to an attitude which encouraged waste, to an excessive application of irrigation water, and to inefficient means of delivery. Concerning the first point, the attitude of the Last Chance organization was described as "the characteristic attitude of the user of water at or near the head of a substantial stream, who until such stream is placed under regulation, takes, because it is there, all of the water flowing in the stream, regardless of its need therefor, or the effect of such taking upon the downstream users."[74] The soils on the Last Chance tract were described as "very shallow,"[75] and, it was argued, that "all authorities agree that on shallow soils light irrigations should be used, not only because they are less wasteful of water, but because they are better for the growing crops."[76] The power company further asserted that it was "incumbent upon every irrigation system to build such reasonably efficient structure, and to establish such reasonably efficient methods of water delivery and use, as to conserve the water for the benefit of all users."[77] Failures of the Last Chance in efficiency of water delivery facilities were illustrated by photographs of some irrigated farmlands south of Grace showing "lava outcrop," ditches filled with moss, evidence of seepage, and lateral canals failing to hold water because of "too low a grade and lack of banks."[78]

The power company attorneys also advanced an innovative but somewhat obscure argument relating to duty of water. Noting that the Bear River began to recede about the first of July, they argued that the duty of water ratio should be adjusted with reference to scarcity during the low water season. Emphasizing the necessity of conserving or using increasingly scarce water economically, they argued that the stage of the river should be considered in fixing an applicable duty of water ratio.[79]

Once the case had been made to the satisfaction of the power company of the water waste in the Last Chance operation, the conclusion was obvious: "The amount of water unnecessarily wasted through the failure to apply such methods (reasonable cultivation and practical methods of applying water) will be deducted in determining the duty of water."[80] The final conclusion, if accepted by the court, would have been devastating to Last Chance interests. "The Last Chance rights proper, cannot exceed a right, with priority as of date of beneficial use of water which was sometime subsequent to May 1, 1902, sufficient to irrigate 15,000 acres of land."[81]

The final consideration in evaluating these appropriations was extent in time. Recall that the purposes of the appropriations included both irrigation and domestic or culinary use. The extent in time of the irrigation season was defined by statute. The extent in time of that portion of the total appropriations applicable to domestic or culinary use was continuous. The problem lay in determining the portion of the total for seasonal irrigation and the portion for continual domestic use. In establishing their argument on this aspect, UP&L cited the experience of both U&I and the West Cache Irrigation Company, large downstream users. In both instances, one-sixth of the summer flow was continued during the winter to provide water for domestic purposes. The power company was prepared to accept this amount "in the absence of evidence" but believed that more was "unjustified."[82]

The cross complaint of the Last Chance Canal Company—that the Bear Lake Reservoir construction by the power company had interrupted the natural drainage from the lake into Bear River—was countered with the position that there had been no natural outlet channel, or at least no channel of any consequence, draining Bear Lake. Whereas the Last Chance testimony had been based on water flow of eighty to one hundred cubic feet per second at the outlet to the river, the power company disputed the source of this water. They maintained that the source was from Paris and Ovid creeks and not from Bear Lake.[83] The conclusion was that the "evidence conclusively establishes the non-contributory character of the lake to the river in the irrigation season."[84]

The Decree

The major event next in the court proceedings following the hearing of June 1919 was the issuance by Judge Dietrich of his memorandum decision, a document he later described as "rather an informal decision but of some length, making a tentative decision as to most of the

claims."[85] This memorandum decision was provided to John F. Mac-
Lane, general attorney for UP&L, who was made responsible for the
preparation of a draft of a proposed decree to conclude the Bear River
Water Case. On May 24, 1920, MacLane forwarded copies of the draft
decree to the attorneys for the various defendants, including J. H.
Peterson, representing the Last Chance Canal Company, for their views.
His letter of transmittal assured the addressees that he had "endeavored
to follow the court's written decision in the description of the rights
accorded to each party, but it is hardly to be hoped that in such an
involved proceeding, I have been entirely successful in avoiding inadver-
tent errors or omissions."[86] With distribution of the draft of the proposed decree, Judge Dietrich ordered responses by all concerned of "objections and exceptions" by the 10th of June. The Last Chance attorney
submitted a list of twelve objections, five of which were accepted in
subsequent coordination.[87]

The court took two actions consistent with the proposed decree before
its issuance in final form. First, on June 24 an order was issued appointing a water commissioner to administer the waters of Bear River and its
tributaries,[88] continuing an office first established in 1918 and still in
existence. And on July 12, under directions of the court, UP&L presented the plaintiff's bill of costs. Included were certain administrative
and printing costs related to the court action that were apportioned at
$4.81 for each of the parties to whom rights were decreed. Costs for the
water commissioner for 1918 and 1919 were also included and apportioned, with the major portions to be borne equally by Last Chance and
UP&L.[89]

The Final Decree, 117 pages long, was signed by Judge Dietrich and
filed with the clerk of the District Court on July 14, 1920.[90] The
document was organized into three sections. Section I comprised general
provisions, limitations, and definitions. Section II accorded decreed
rights to four hundred parties to the case.[91] Section III listed those
defendants receiving no decreed rights and included a statement of areas
in which the court retained jurisdiction.

That the Bear River was "an inter-state stream with a large number of
tributaries in the States of Utah, Wyoming and Idaho" was recognized.
Only that portion of the river from the location of the Stewart Dam at
the point of diversion into the Rainbow Canal downstream to the Utah-
Idaho border, together with all intervening tributaries of the Bear River,
was drawn within the jurisdiction of the District Court, District of

Idaho, Eastern Division, for purposes of the case.[92] As a partial exception, however, certain rights decreed to UP&L and U&I were included in the Schedule of Rights even though the points of diversion for these waters were in Utah. Emphasis was provided to the assertion that this was no adjudication of title to these rights, but was merely "a recognition of said rights to the extent that in the administration of that part of the river within the jurisdiction of this court, . . . shall see that there is delivered at the Utah state line such quantity of water as is necessary . . . to satisfy said rights in accordance with their dignity and priority as herein recognized."[93]

A general analysis of Section II, "Schedule of Rights," shows that a total of 9,212.3 c.f.s. were decreed for beneficial use for power and irrigation. In addition, storage rights for 6,000 c.f.s. were decreed for the Bear Lake Reservoir and 4,000 acre-feet for storage in a small reservoir at the Cove plant, below Grace. Of the total water decreed for power and irrigation, 83 percent was from the main river and 17 percent was from sixty-three tributaries (creeks and springs) of the Bear River. Priorities were early on the tributaries. Five creeks had appropriations with priorities of 1864, 1868, 1869, and 1870. Priorities were later on the main river, with the earliest priority decreed for 1879. During the period 1879 to 1883, six appropriations were made, totalling only 35.2 c.f.s. Although important to individual water appropriators on the tributaries, such water was of no substantial interest to the plaintiff, UP&L.[94] That portion of the decree pertaining to tributary water is therefore omitted from further analysis or discussion here. Of the 7,647 c.f.s. of main river water for power and irrigation covered by the decree, 79 percent was for power and 21 percent for irrigation.

It would appear that the decree accorded to the Last Chance Canal Company and to its subsidiary corporations most of what could logically have been expected.[95] True, their vague claim to 6,616 c.f.s. of Bear River water from predecessors in interest was disallowed. But in adjudicating priorities of appropriations, the court accepted the doctrine of relation so that priorities were determined from the dates of posting of the notices of appropriation. For both the 1897 and 1901 filings by Last Chance, the dates the documents were presented for notarizing were accepted by the court. The extents in quantity of the Last Chance claims as decreed were: March 1, 1897, 200 c.f.s.; and May 14, 1901, 240 c.f.s. The decrease in appropriations claimed of 1,000 c.f.s. to 440 c.f.s. was an accurate reflection of the capacity of the diversion works constructed. There was no obvious explanation of why the decree associated 200 c.f.s. with the 1897 filing and 240 c.f.s. with that of 1901. The separate claims by the Bench Canal Company with priority dates of August 9 (138.16 c.f.s.) and December 31, 1909 (25.6 c.f.s.), and by the Tanner Canal Company for 54 c.f.s. with a priority of July 29, 1910, were decreed without change. Apparently, the lengthy arguments by UP&L concerning duty of water had no effect in reducing claims of elements of the Last Chance system. However, a general admonishment for economy did appear. "Notwithstanding this schedule of Rights, users of water under this decree shall at no time divert more water than can be beneficially used, and waste of water is hereby prohibited and enjoined."[96]

The court's adjudication of Last Chance claims in extent in time was more complex. The irrigation season was defined as "that portion of each calendar year which commences on the 20th of April and closes on the 30th day of September." However, during the periods April 20 to 30 and September 15 to 30, "no irrigation appropriator shall divert or use more than 40 per cent of . . . allotment under the 'Schedule of Rights' hereinafter prescribed." Further, each irrigation right "shall include and imply as a part thereof a domestic right to the use, during the non-irrigating season, of such waters allotted for irrigation purposes as are necessary for . . . domestic purposes."[97] Unfortunately, sufficient evidence was not considered to be available so that the extent of the use of water through irrigation canals in the nonirrigation season could be "definitely fixed."[98] The court, therefore, retained jurisdiction "to take further evidence upon and to increase or decrease . . . allottments {*sic*} for winter use." Jurisdiction was also retained "to review and amend the provisions of this decree, fixing the limits of the irrigation season, and further reducing the amounts of water which may be diverted for irrigation purposes during the months of August and September upon proof of decreased requirements by any appropriator."[99] The reservations of jurisdiction were to end with any later term of the court before January 1, 1924, which "may determine these matters."[100]

The cross complaint of the Last Chance Canal Company, alleging interference by reservoir construction at Bear Lake with natural flow from the lake into Bear River, brought a measure of success. As full "compensation for any interruption of the natural flow from . . . Bear Lake area by the plaintiff's reservoir works," the Utah Power and Light

Company was ordered to discharge annually the following quantities of water as "natural flow" from the lake into Bear River: April 20 to July 1, 50 c.f.s.; July 1 to 15, 35 c.f.s.; July 16 to August 1, 25 c.f.s.; and August 1 to September 15, 15 c.f.s.[101]

The Dietrich decree gave UP&L both storage and power rights. Appropriations in 1902 were made for 2,000 c.f.s. (March 24) and 3,000 c.f.s. (April 12) for storage of Bear River water in Bear Lake, to be used as a reservoir. Another appropriation for 2,500 c.f.s. for storage of Bear River water, plus 500 c.f.s. from Bear Lake and Mud Lake, was made on September 11, 1912. This last appropriation occurred about a year after the reservoir construction had been completed sufficiently for diversion, storage, and release of stored water. The decree bestowed storage rights for 5,500 c.f.s. of Bear River water, 3,000 c.f.s. with a priority of March 1, 1911, and 2,500 c.f.s. with a priority of September 11, 1912. The priority for storage rights to the 500 c.f.s. draining naturally into Bear or Mud lakes was September 1, 1912.[102] Concerning diversion from the main river for storage, the appropriations of September 1912 were decreed without problems. However, only 3,000 c.f.s. of the total of 5,000 c.f.s. appropriated in 1902 were decreed. Bestowal of rights was, presumably, contingent on completion of the reservoir project for actual use and the capacity of the diversion works as they existed in 1911.

To accomplish the impounding and storage of water from the Bear River, the diversion was authorized "at all times, and at all seasons of the year." Diversion and impounding of "all of the waters of Bear River to the extent of 5500 cubic feet per second of time, together with the waters naturally flowing into or arising in said lakes" was approved.[103] Authorization for diversion of the entire flow of the river was conditional and subject to the proviso that the power company discharge "at the same time through its outlet control works . . . when there is need therefor to supply the rights of prior appropriators below . . . an equivalent amount of water, such quantity to be regarded as natural flow of the river, and not released stored water."[104] Such waters stored could thereafter be "released at the plaintiff's pleasure" for return to the natural channel of the river "for use at various points of diversion now existing, or which may hereafter be established by the plaintiff for the generation of electric power, and for such irrigation or other beneficial purposes . . . as the plaintiff may devote or dedicate said released stored waters, by use, sale, rental, or otherwise." Water, discharged into the natural channel of Bear River, was "protected . . . for the distribution designated by the plaintiff, as though kept and conveyed within an artificial channel."[105]

These storage rights, with the adjudicated priorities, were "subject to the prior rights of the various defendants, as hereinafter decreed." Use of these rights was not "to interfere with the exercise of any prior rights fixed by this decree."[106] These prior-rights considerations were important. Of 1,607.76 c.f.s. of main river water for irrigation, all but 43 c.f.s. had priorities earlier than those for the storage rights of UP&L.[107]

Two other categories of rights were associated with the hydroelectric developments on the Bear River. First, storage rights for 4,000 acre-feet of Bear River water were decreed for the Cove plant in 1916.[108] Second, power rights, continuous throughout the year, were decreed for a total of 6,040 c.f.s. of main river water.[109]

Evaluation and Subsequent Litigation

Six months after the Dietrich decree was issued, the water commissioner, who had administered the distribution of water from the Bear River since 1918, was able to report, "great satisfaction has been expressed over the decree, and practically every water user considers the problem is finally settled."[110] However, two issues remained potentially unresolved that were subjects not adjudicated by the final decree and over which the court had retained jurisdiction. A commitment to accumulate additional evidence for decisions before January 1, 1924, was made.

The first subject requiring further study and decision was fixing the extent of the use of water during the nonirrigation season for domestic purposes. Only one instance of addressing this aspect appeared in the official record. In November 1923 the Last Chance Canal Company appointed two individuals, either of whom was authorized "to get with the water commissioner to determine the amount of water needed in our canals for culinary purposes."[111] Current research has not disclosed any additional actions to resolve this problem. Water use for domestic purposes was so inconsequential that neither rights nor measurements were available or required. Some canals of the Last Chance system carry water throughout the year for use by livestock; others close completely as soon as heavy frost occurs.[112]

More troublesome was the second subject—a retention by the court of jurisdiction to review that portion of the decree fixing the limits of the irrigation season and to consider further reduction in the quantities of water decreed for irrigation during the months of August and September. Anticipating the problem, W. O. Creer, a prominent Last Chance stockholder, urged an aggressive policy to "get the water in our canals on April 20th" to circumvent any suggestion to Judge Dietrich that the irrigation season could be shortened. Creer deplored being considered a "calamity howler" but was firm in his intention to "get the full amount of water decreed and also [to see] that the irrigation season be firmly established to the full limit."[113] By October 1922 the Last Chance's official position, to be expressed in a petition supported by stockholders' affidavits to the court, was to obtain a change in the decree "so that we may have our full stream until October 1 of each year."[114] Water was believed to be needed during this period, particularly for irrigating sugar beets, alfalfa hay (12,800 acres), and pasture (3,200 acres).[115]

On September 15, 1923,[116] the Last Chance Canal Company was notified that the reduction in the water authorized for irrigation before April 30 and after September 15, as specified in the 1920 decree, would be made final. The Last Chance fought bitterly this tentative decision to decree water at 40 percent of that listed in the Schedule of Rights for these two short periods at the beginning and end of the irrigation season. The canal company carried out an expensive and unproductive approach of soliciting support from the Federal Water Commission. The court remained unconvinced of the need for change and the 1920 ruling became final.[117]

Probably symbolic of the reaction of Last Chance officials and stockholders to this nearly seven years of expensive and sometimes bitter litigation was that in November 1924, Last Chance had not yet paid the cost of $1,030.01 to UP&L, which was properly charged as incident to prosecution of the Bear River Water Case.[118] Defeat was not palatable to these early farmers of Gem Valley. Any form of surrender was never easy.

The Bear River Water Case was early recognized as one of the potential milestones in the irrigation experience in Idaho. The structure of the court case, a summary of the contending legal arguments, and a discussion of the terminating decree and certain other subsequent developments have been the subjects of this chapter. However, the effects of the Bear River Water Case also were to be felt in a much broader context, that is, in a manner significantly influential on Western water policy. These aspects of the Bear River Water Case constitute a major portion of chapter nine.

CHAPTER EIGHT

Challenges
and Solutions

During the half century since conclusion of the Bear River Water Case, many problems have confronted the farmers of the Last Chance canal system. However, among all these perplexing circumstances, two categories in particular stand out as most important. One was a remnant of all their history—a continuing concern for adequate water for irrigation. The second involved a decision to expand activities to include the generation of hydroelectric power.

The Search for a Guarantee of Adequate Water

The final adjudication of the waters of the Bear River covered by the Dietrich decree did not, unfortunately, provide a final, consistent solution either to the power or irrigation interests along the river. Perhaps most important, the influences of weather and climate proved to be unpredictable, irregular, and, in many cases, dominant.

To consider the effects of the weather and climate on potential water availability to various irrigators, several features of Idaho water law should be recalled. Waters subject to appropriation were those flowing in the natural watercourse of a stream, the so-called "natural flow." The priorities of the various appropriations were "first in time, first in right." This meant that the prior appropriator was entitled to sufficient water at his point of diversion to supply his right. If there was insufficient natural flow to provide enough water to all appropriators covered by the decree, appropriators were denied water in the inverse order of their priorities.

The Dietrich decree covered slightly over 7,647 c.f.s. of the natural flow of the Bear River for both power and irrigation, excluding tributaries that are not of direct concern to this study. Among the appropriators, the Last Chance Canal Company and its subsidiary companies were accorded high priorities in time. Even the Last Chance system appropriation "youngest" in time—that of 54 c.f.s. decreed to the Tanner Canal Company with a priority date of July 29, 1910—was in the top 40 percent in priority of all decree appropriations. By contrast, priorities decreed to UP&L were extremely low. Only 43 c.f.s. of water decreed for irrigation had a lower priority. The power company was, however, authorized to divert all the natural flow of the Bear River, up to 5,500 c.f.s., from Stewart Dam, through the Dingle or Rainbow inlet channels, and into Bear Lake, provided that, when necessary to meet the rights of prior appropriators, there was discharged through the outlet control works an amount of water equivalent to the natural flow of the river. This "redirected" natural flow was to be further augmented by amounts varying from 50 c.f.s. to 15 c.f.s. during the period April 20 to September 15 to compensate for the interruption to natural flow from Bear Lake caused by the power company's reservoir dikes. Waters diverted into Bear Lake and not returned to the river as natural flow were considered as stored waters that were available for release, at the pleasure of the power company, for power, irrigation, or other beneficial use. Use, sale, and rental of these stored waters were specifically authorized.

Thus, it becomes clear that the amount of water available at any given time for a specific irrigation appropriator would be controlled by the quantity of water decreed to that irrigator, the priority in time, and the quantity of natural flow of the river. In many respects, this latter feature assumed a position of major influence. Experience data from measuring, for three years, the natural flow of the Bear River at a point in Idaho upstream from the Stewart Dam and near the Idaho-Wyoming border illustrates this point. Russell D. Stoker, the Bear River watermaster since 1954, provided a tongue-in-cheek description of 1964 as an "average" year. On June 10, 1964 the natural flow was 2,140 s.f. In 1983, an extremely wet year, the flow measured at 5,460 s.f. on this same day. By contrast, 1977 was one of the driest years on record. On June 10 the natural flow was only 88 s.f.[1] The vital importance of reservoir-stored water for circumstances such as occurred in 1977 is all too clear.

Considering the general thoroughness of the Last Chance planning for their irrigation operations, their apparent lack of early concern for a reservoir reserve is difficult to understand. During the court proceedings incident to the Bear River Water Case, Last Chance officials spoke disparagingly of UP&L and its Bear Lake Reservoir. Circumstances, however, quickly demonstrated the extent to which the Last Chance system irrigators were dependent on water from that source. Such dependence occurred commonly in the latter part of the irrigation season during dry years.[2]

Within ten years of completion of the minimum essential reservoir works, the Last Chance system was renting supplemental irrigation water from UP&L—the first time being in 1919. During the next sixty-two years (1919–80), there were only fifteen years when it was not necessary to rent water from UP&L. Annual rentals averaged 6,390 acre-feet during that period.[3] Rentals in excess of 10,000 acre-feet were required fourteen times; rentals in excess of 20,000 acre-feet were required seven times; and in 1961 a record-high 43,973 acre-feet were rented—fully 45 percent of the total water used.[4] The average annual supplement required was almost 7 percent.

The year 1934 illustrates this criticality of water availability in more subjective terms. Last Chance officials described the availability of rental waters as "our only salvation."[5] And UP&L was not reluctant to take the credit. With justifiable pride, their officials reported the results of their having provided rental water, to include 25,188 acre-feet to the Last Chance system. "Farmers of Utah and Idaho along the Bear River saved crops valued at $3,832,906 from destruction in 1934 because pumping on Bear Lake made water available for the drouth stricken areas along the river."[6]

Even though the needs of the Last Chance system for supplemental irrigation water from the Bear Lake Reservoir are easily documented, dry years, particularly several in succession, produced problems. By design the reservoir impounded water at elevations between 5,922.65 feet and 5,902.00 feet. The reservoir water thus measured 20.65 feet vertically. When full, the top 14 percent of the water stored could be released by gravity flow; the remainder had to be pumped by the Lifton plant. Even the pumping of water from the reservoir was not without problems. In 1977, during pumping operations it was discovered that a large sand bar blocked access to the pumps when the water level was down to about 5,914 feet. Extensive and expensive dredging was required so that the water could reach the pumps.[7]

During the history of the Last Chance system several severe droughts have occurred, causing low reservoir water levels as well as decreased natural river flow. In these instances, Last Chance was desperate for water to save crops and UP&L had to tap deeply its lake reserves by pumping. For example, the following reservoir elevations were recorded: 1930s, 5,902.00 feet (reservoir depleted);[8] July 10, 1934, 5,907.46

feet;[9] 1961, 5,911.00 feet; 1977, 5,914.09 feet.[10] Under these circumstances the needs of Last Chance and other irrigators were critical. But
other interests vigorously opposed further reductions in the elevation of
Bear Lake owing to pumping operations by the Lifton plant. As the
water level receded, some boating and resort commercial interests along
the lake shore became isolated. Serious agitation and mass meetings
opposing further "withdrawal of stored water from Bear Lake" were
subjects for attention by the governor of Idaho.[11]

Reflecting perhaps a certain lack of confidence in the future capability of the Bear Lake Reservoir to meet consistently every year the
supplemental irrigation requirements of the Last Chance Canal Company, alternate solutions were sought. Last Chance, in conjunction
primarily with UP&L, expressed planning interest in a new dam on the
Bear River near Soda Springs. This dam would create a new, closer
reservoir having a capacity of 40,000 acre-feet. The power company
visualized use of the major portion of the stored water for cooling a
thermal power plant possibly to be constructed in the vicinity. Under
the plan, the Last Chance Canal Company would have 6,360 acre-feet
of the total, based on their sixty-two year average requirement for
supplemental irrigation.[12]

The Last Chance Hydro Electric Company

In the 1980s Last Chance became actively involved in an entirely
different enterprise—the generation of electric power. Ironically, this
involvement would lead to apparent solution of their longstanding problem, the insuring of enough water for irrigation.

The experiences with electricity of the inhabitants of Gem Valley,
particularly of the farmers served by the Last Chance canal system, were
meager during the early years of this century. The first hydroelectric
plant on the Bear River, completed at Grace in 1908, provided power
for local domestic consumption in 1909. But this occurred only after the
formation of a local business organization to buy power wholesale from
the power company plant, reduce the voltage with transformers, and
market the electricity to local consumers in the village of Grace.[13]
However, farms in the valley were not provided with electricity. To
eliminate this deficiency, Last Chance attempted an involvement directly
in the generation of electricity through use of water from the Bear River.
In late 1920 the canal company considered a proposal to change its

articles of incorporation so that power rights might be held.[14] The company acquired a power site on the river but was unsuccessful in its attempt to file on water for generating electricity—this for the reason that there was no more water available for that purpose.[15] In early 1925 action was being taken with their attorney, J. H. Peterson, to determine how to dispose of the power site and related filings controlled by Last Chance.[16] The mid-1930s brought the Rural Electrification Administration to the national scene and with it a policy that "power must be extended wherever it was demanded."[17] Implementation of this policy within Gem Valley ended any immediate requirement that Last Chance become further interested in the generation of power. The matter rested for over forty years.

By 1980 circumstances had changed. "The shareholder-farmers in the Company had suffered several increases in the cost of electrical energy during previous years and were interested in developing their own source of electricity to off-set these costs."[18] UP&L constituted the "most practical market" for any power produced. In December 1980, this market was assured when the Idaho Public Utilities Commission Order Number 16048 "required The Utah Power and Light Company to purchase electrical energy from independent power producers."[19] An aura of destiny marked many of the activities of the Last Chance Canal Company, and this was true in their preliminary consideration of a hydroelectric project in 1980. The theme, fulfillment of a dream, was seen in the possibilities of a hydroelectric project. Fred Van Vleet, Last Chance watermaster for many years, long dreamed of the feasibility of such a project and, indeed, he had tentatively selected the exact location for such a power generation facility.[20]

The first steps in converting the possibilities of a Last Chance hydroelectric project into reality were investigative and administrative in nature. In early April 1980 came the first formal expressions of interest by Last Chance officials in such a project.[21] By April 19 the canal company directors had decided to pay J-U-B Engineers of Pocatello $1,600 for a site reconnaissance study. Such a study was required to accompany an application to the U.S. Department of Energy for a loan to cover expenses of an evaluation, or feasibility study, to determine the real potential for a hydroelectric generating facility.[22] On May 5 further contractual arrangements were made with J-U-B Engineers to aid in the preparation of the loan application. Further, as a contingency upon acceptance of the application by the Department of Energy, J-U-B Engineers was to initiate the feasibility study itself.[23] In the meantime the Last Chance attorney filed an application for permit to appropriate the public waters of the state of Idaho. This application, dated May 30, 1980, requested 440 c.f.s. of water from Bear River for yearlong nonconsumptive use for power purposes. This application was approved by the director, Idaho Department of Water Resources, on December 18 with a priority as of the date of filing.[24] Two aspects of this filing are of interest. First, the extent in quantity was equal to that covered by the Dietrich decree of 1920. Second, the approval, in effect, vacated limitations of the Dietrich decree concerning restriction to 40 percent of the adjudicated water appropriations at the beginning and end of the irrigation season and the limitation to water solely for culinary or domestic purposes during the nonirrigation season.

On September 2, 1980, feasibility study funding was approved by the Department of Energy for a total of $31,872. This loan was made under provisions of the "forgivable loan" program. Under this program 90 percent of the funding came from the Department of Energy and 10 percent from Last Chance, with the provision that if the project under consideration was judged infeasible, the loan was "forgiven." But if the proposed project was judged feasible, the Department of Energy loan was assessed interest at 7.5 percent annually and was to be repaid within ten years.[25]

Approval of the application for the loan for the feasibility study on September 2, 1980, brought quick action by J-U-B Engineers. The "Feasibility Study for the Last Chance Canal Company . . . to the U.S. Department of Energy, Small Hydroelectric Project" was published in March 1981. This study was predicated on a series of principles or "design considerations,"[26] several of which are particularly pertinent to this discussion. "Power generation equipment, structures and appurtenances should not interfere with normal irrigation practices and required flows." Although in concept the project visualized generation of power primarily during the nonirrigation season, plans also included use during the irrigation season of some water to operate the turbines, return of that water to the river, and recovery of that water downstream by the Bench "B" Canal for use in irrigation.[27] Two other principles were important. "Existing structures and features should be utilized wherever possible." And, "the most cost-effective alternative should be given precedence."

Application of these principles resulted in a plan that called for utilization of the water appropriated for power purposes and the existing

Last Chance diversion dam, control headgates, and main canal to a point to be selected for new facilities associated with the hydroelectric project.[28] Three alternate locations on the main canal were considered for the new facilities. One location, situated about one hundred feet upstream from the tunnel inlet, was recommended for development. This location offered fewer design, construction, and operation problems. Development there was less expensive and would provide similar revenues.[29] In summary, waters were to be diverted from the Bear River as for irrigation, passed through the main canal to the point selected for the new hydroelectric facilities, and, through penstocks with a drop of slightly over thirty feet, passed through the facility's turbines before the stream diverted for power purposes was returned to the Bear River.[30]

The plan for the hydroelectric project involved some improvements to existing facilities such as changing the control works to automatic devices and improving the main canal for maximum water flow. Provision of new facilities, however, constituted the major portion of the plan. A new access road and a bridge over the main canal were required, as were transmission lines to carry the electricity from the generator plant to a connection with UP&L lines. At the location selected in the main canal, a penstock diversion structure was required. Associated with this large concrete box was automatic-flow movement and control equipment, which insured the proper flow of water into the tunnel for the irrigation system and the diversion of the remainder into the penstocks for delivery to the turbines. A building to contain the turbines, generators, and some control equipment was to be located almost at river level below the penstock diversion structure. From the level of the main canal, an emergency spillway was also planned for release of water during periods when the turbines were not operating.[31] A computer building was also planned near the highway just north of Grace. (See map on page 44.)

Conclusions based on this feasibility plan were that this was a "financially feasible project" estimated to cost $1,428,600. Power sales for the first year were estimated to be $400,450.[32]

Even before the feasibility study was published, certain related activities were in progress. During the fall of 1980, Last Chance instituted a publicity effort to acquaint all interested parties with details of the hydroelectric project as they developed.[33] This effort was successful; on February 16, 1981, the stockholders of the canal company voted unanimously "to build a hydroelectric power plant on their holdings on the north end of the canal tunnel." Associated problems included acquisition of financing for the necessary 1.4 million dollars at "the best interest rate" and putting the "construction out for bids."[34]

In January 1981 the canal company made a second filing for Bear River water for power purposes. In this instance the appropriation sought was an additional yearlong 220 c.f.s., thus giving total appropriation for power purposes of 660 c.f.s. The primary reason advanced for the second filing was that "a preliminary feasibility study . . . determined that a different place of use up stream was more feasible which would permit utilization of larger quantities of water."[35] This application for water was approved by the Idaho Department of Water Resources in mid-year.[36]

A new subsidiary corporation, The Last Chance Hydro Electric Company, was established on June 19, 1981, to construct and operate the planned facility. This new company maintained the nonprofit status of the canal company.[37]

Financing of the project was arranged through the Spokane Bank of Cooperatives. On June 4, 1981, two loan applications were approved— $1,540,000 for project construction and $225,000 for project expenses already incurred. The loans, having a "floating" interest rate then at 13¼ percent, were to be repaid within fifteen years beginning when the plant went into operation, a date projected for January 1983.[38]

Several major firms were involved in construction. J-U-B Engineers had primary design responsibility. Shunn Construction, Incorporated, Ontario, Oregon, received the contract for the access roads, bridge, diversion works, and penstock stubs. Ingersoll-Rand Company of Boston supplied the three turbines/generators. Installation was by Westinghouse Company of Pocatello. The computer-building contractor was Allan Horsley.[39] Last Chance directors exercised much personal supervision over the various construction activities.

Some improvements to existing facilities were made during the fall of 1981.[40] Construction of the new facilities began on March 23, 1982.[41] This was arbitrarily described as Phase I and included all the work contracted to Shunn Construction. This phase was to be completed by May 10 so that this work would not interfere with the start of the irrigation season.[42] The turbines/generators were scheduled for delivery in October and installation in November 1982.[43] Work started on the computer building in late June 1982 and was scheduled for completion by August 1.[44] The transmission lines were completed and ready for connection at both ends.[45]

On January 15, 1983, with two of the three turbines/generators producing power, the Last Chance hydroelectric project was operational.
The operational status of the third turbine/generator was delayed because
of a "hot bearing." Cost of the project was $2,176,000, 55 percent over
estimate.[46] The capacity of the system was rated at 1.4 megawatts.[47]
What had been built was a "totally computerized facility"—a "showcase
for small hydroelectric projects."[48]

In an action reminiscent of one in 1912 when UP&L removed U&I
from the role of power generation on the Bear River, the power company
initiated action to acquire the Last Chance Hydro Electric Plant before
its completion.[49] In a communication to Last Chance discussing their
demonstrated needs for supplemental irrigation water, the power company summarized three possible alternatives. One was for Last Chance to
continue to obtain supplemental water under the present contract of
August 20, 1979. However, Last Chance was already aware of a need for
different arrangements for supplemental water. Costs were becoming so
prohibitive that they might "force a change in our present farming
practices and crops raised in the future." Another alternative was for Last
Chance to pay those costs for the Soda dam, if eventually built, related
specifically to the storage of an additional 6,360 acre-feet of water to be
used as supplemental water by the Last Chance system. This 6,360
acre-feet value was, it will be recalled, the average annual requirement of
the Last Chance system for rental waters.

The third alternative concerning continued availability of supplemental water for irrigation involved sale of the Last Chance Hydro Electric
Plant to UP&L. Under this alternative, in return for full title to the
plant, UP&L would, first, pay all the loans and other expenses attribut-
able to the plant construction (about 2.2 million dollars). Second, the
power company would provide permanently at no expense to the Last
Chance system supplemental irrigation water as needed from Bear Lake,
or from the reservoir established from the Soda dam, if built. Amounts
of water to be provided were to be established at an average of 6,360
acre-feet per year. For dry years, up to 20,000 acre-feet annually could
be required, provided a total of 63,600 acre-feet was not exceeded over a
consecutive ten-year period. Water in excess of these values would be
provided at current contract costs by the power company if the need
arose and water was available in the Bear Lake Reservoir. Finally, with
the so-called "fallen water charge," a form of royalty, the power company would pay to Last Chance five mills per kilowatt hour of electricity

produced by the plant. This payment would be tied to the consumer index and could be adjusted upward as the index indicated.

In August 1982 the Last Chance directors voted to negotiate with UP&L concerning their offer to buy the "hydro-electric plant which is now under construction." The reason for this action was stated to be "an effort to obtain more water for the Last Chance canal system."[50] It is not clear what the "negotiations" accomplished. The sale, accepting the earlier offer of UP&L, was approved unanimously by Last Chance stockholders on December 13, 1983—eleven months after the new plant officially became operational.[51] The sale contract was signed on January 19, 1984, by representatives of both UP&L and Last Chance.[52] The power company took over operation of the hydroelectric plant on that date, although final approval of the sale was contingent upon action by a "myriad of government agencies."

The Last Chance Hydro Electric Plant marked the "first time any such undertaking had been accomplished in Idaho."[53] After successful realization of the initial project goals, why did Last Chance sell? Other than the obvious benefits of royalty payments, relief from debt, and permanent acquisition of supplemental irrigation water, there were other considerations. First, the canal company had operated the plant for one year and had on hand about $100,000 in profits, which would be retained. Second, the cost overrun had increased payments so much that little profit could be expected in the foreseeable future. Third, there were other risks that were worrisome. An especially dry year was always possible wherein insufficient water to operate all three turbines/generators would cause curtailment of production and subsequent loss of revenue. And, of course, there was always the risk of catastrophic accident to the plant.[54]

Regardless of the reasons for the sale, however, this venture by the Last Chance organization must rank as one of the ironies of the history of these people and the area in which they live. For almost a century the primary concern was the availability of enough water for irrigation of their lands. A principal competitor for this water was the Utah Power and Light Company. Possible solutions to the problem had explored diverse paths. Building a hydroelectric plant would finally achieve the goal of contracted guarantee of the availability of irrigation water to meet their needs.

CHAPTER NINE

An Assessment

Before the word *finis* can be affixed to the bottom of the last page of this story of the Last Chance irrigation system and of certain closely associated external developments, it is appropriate to assess the significance of the events illustrating the story. This assessment follows a generally chronological order and may partially reflect the achievements of significance to the irrigation project itself or to the communities and valley it served.

However, if the experience of the Last Chance Canal Company can be considered a case study, a microcosm, in frontier irrigation history, a premise first stated in the introduction, events and patterns associated with it should be discernible as having influenced other, more comprehensive developments. A true appreciation of the success of the Last Chance enterprise and an understanding of the role of the associated external developments requires that both be judged not only for their local implications, but also for connotations in a broader, historical perspective. Developments in this latter category become illustrative of Western water policy.

Regrettably, the historical evidence available does not support clear cause-and-effect arguments and conclusions. In contrast to the easily identifiable and well demonstrated local accomplishments, the broader aspects are much more difficult to establish and explain. Nevertheless, the logic of an association of events and subsequent related developments is apparent and must be recognized. Events cited which follow in time from Last Chance efforts and accomplishments almost certainly can be attributed, at least in part, to the earlier initiatives by that company. Moreover, the fact that Last Chance was a protagonist in the Bear River Water Case, one of the most important water rights adjudication cases of the early twentieth century, enhances its position as an influential factor in early Western water policy.

The goal of the Last Chance Canal Company was entirely self-oriented and single-minded: to irrigate their lands. There was no awareness of the need for, or interest in, the establishment of broad policy. Policy, however, would eventually evolve irrespective of this lack of concern. The essence of a series of events, occurring primarily over the first quarter of this century, provided the bases for the policy we seek.

In retrospect, it appears that much of the knowledge and many of the activities of those who contributed to the Last Chance effort, while commendable in terms of pioneer traditions and achievements, are not unique. Some prospective settlers arrived on the Gem Valley portion of the Bear River in 1895 strongly imbued with the need for irrigation as prerequisite to successful settlement. John Trappett, one of those settlers, recalled: "We prospected up the river to see what show there would be of getting the water out before we went to Blackfoot [the federal land office] to file on our homesteads."[1] Since these settlers emigrated largely from Mormon settlements south of Gem Valley, they were also aware of Mormon irrigation techniques suitable for use in the new settlements. However, the extent of local, active church involvement in the Last Chance project is not clear. When the project plans were first being formulated, Trappett spoke of a "representative from each ward,"[2] but no further mention of church participation appears.

As a final consideration in establishing parameters for the forthcoming Gem Valley settlement-irrigation effort, it is noteworthy that Idaho water law was in existence at the time of settlement and had remained virtually unchanged since territorial days. Thus, from a background of known quantities, the settlement and irrigation of Gem Valley proceeded.

The construction phase for the Last Chance irrigation system was, in many respects, one of the most dramatic and unusual aspects, particularly as compared to other Idaho irrigation developments. The engineering challenges incident to diversion of waters from a canyon of the Bear River onto lands many miles from the point of diversion were great. But even after over a decade of failures fresh in mind, these indomitable pioneers were not in awe of the complexity of their task. Irrigation was essential and irrigation there would be, whatever the costs in time, labor, and money.

An attitude of independence and resourcefulness true to the best of American traditions was particularly apparent in the way the Last Chance project was financed. In a period in U.S. reclamation history when outside financial assistance was generally the rule, the Last Chance builders opted for locally arranged, private funding. As expressed by an early official of the canal company: "This Last Chance Canal Company, Limited, has become famous throughout the west because of the fact that it was built without federal assistance, and also without the outside capital that usually is necessary in undertakings of this magnitude."[3] This concept of financing was in marked contrast to methods followed in other contemporary irrigation developments in Idaho. For example, the New Plymouth development of 1885 in the Payette Valley was an "irrigation colony" financed by eastern speculation capital.[4] Benefits from federal reclamation legislation brought the Twin Falls Land and Water Company into existence in 1903 as the largest Carey Act project in the nation.[5] The large Minidoka Project, also on the Snake River, was begun in 1910 under the auspices of the Newlands Act.[6] That these latter two reclamation projects were both larger and more expensive than the Last Chance system does not diminish the aura of Last Chance perseverance—a proof that irrigation can be achieved "on your own," if both the need and the will are great.

Another noteworthy feature of the construction phase relates to the dangers inherent in this type of work performed by men unskilled in such activity and using only primitive tools and machines. Regardless of the tremendous risks involved, there was only one casualty during the entire project. Tab Merrill was killed by a falling rock early in the construction.[7] This was a remarkable record in safety. But retention of the Tab Merrill tale in Gem Valley folklore long after the event is even more significant. Obviously, a genuine concern existed for the individuals who performed this work.

It is puzzling that in the overall visualization of the Last Chance irrigation system, there was no provision for reservoir facilities. Although both Powell and Chittenden had espoused a federal interest in Bear Lake as a reclamation reservoir, Last Chance during the early years retained its focus solely on the natural flow of the river as the source of irrigation water.

The Last Chance irrigation system became operational in 1902 when the first diverted Bear River waters flowed from the canal onto farmlands on the east side of the river, and irrigation at long last became a *fait accompli*. For over eight decades the Last Chance has continued as a successful and expanding enterprise, an enterprise that eventually provided irrigation to over thirty thousand acres. The significance of successful completion of the Last Chance system was obvious to Gem Valley residents. Capability to irrigate brought a sense of permanence for the settlers. With irrigation came prosperous farms, comfortable homes, better schools, and justification for a new church.[8] Gem Valley became a flourishing example of successful settlement.

Claims for some of the last good agricultural lands of the Bear River watershed and for Bear River water had brought success to the settlers. Potential competing interests saw other values in the Bear River water, and the focus of this competition came at the village of Grace, the location of the Last Chance Canal Company. There in the river canyon, a fall of 525 feet dazzled the hydroelectric developers, and they lost no time in planning for construction. The appeal of the Grace location for the first hydroelectric plant on the Bear River is highlighted by the fact that the second-best location, Oneida, has a fall of only 145 feet, only 28 percent of the potential at Grace.[9] The plant at Grace became operational in 1908. The development of the Bear Lake Reservoir and the "entire river concept" held by UP&L quickly followed. With these developments "the Bear River became one of the most scientifically controlled streams for irrigation and power in the United States."[10]

As competition for Bear River water among the various competing irrigation and power claimants intensified, it became necessary for the courts to establish a legal basis for distribution. For Last Chance, among others, and UP&L, this was accomplished by the Bear River Water Case in litigation from 1917 to 1920. In this case, Last Chance and its subsidiary canal companies were identified as the "principal adversaries" of the power interests. The case, one of the earliest court actions of this type in Idaho, concluded with a sound, fair verdict from an eminent jurist. His decision and decree became legal precedent and gave meaning in practice to the theory of Idaho water law.

Basic to the Dietrich decree of July 14, 1920,[11] court jurisdiction was applied only to that portion of Bear River from Stewart Dam, the primary point of diversion of Bear River water into Bear Lake, downstream to the Idaho-Utah border. For this portion of the Bear River, the difficult question of the extents of the various appropriators' rights in quantity were adjudicated. The decree allocated over 9,200 c.f.s. of water for both irrigation and power. Of this total, 83 percent was from main river water and the remainder from the various tributaries of the Bear River. Concerning the adjudication of rights to the main river water, solely those of interest to either the power company or the Last Chance irrigation system, 79 percent was for power and 21 percent was for irrigation. Of the total of slightly over 1,600 c.f.s. of main river water adjudicated for irrigation, almost 41 percent represented the rights in quantity of the Last Chance system.

Other provisions of the Dietrich decree had effects or implications of a much broader nature than the adjudication of water rights. Two of these additional, important features were the legal consideration of a highly theoretical concept of "duty of water" and a recognition of the interstate nature of the Bear River. From this last concept came the Bear River Compact and the Bear River Commission, now respected features of interstate river management. Also, the Bear River Water Case became important as a legal precedent in other court actions or in governmental considerations. Each of these subjects, discussed below, had Last Chance association and significant water policy implications.

The concept of "duty of water" as a measure of the genuine, justifiable needs of the appropriator was considered as applying to determination of the extent of a right in quantity. This rather nebulous concept for economy and reasonableness in use of water had been assigned a tentative, "normal" value in 1903 of one cubic foot per second of water being appropriate to irrigate fifty acres.[12] The power company argued at great length for increasing the number of acres assumed capable of being irrigated by one cubic foot per second of water—and thereby reducing the competing irrigation rights in quantity. Irrigation interests wanted to dispose of this matter with a general statement reflecting a belief that the land requirements for water varied extensively and were so unpredictable that they defied assignment of precise values. Judge Dietrich avoided this quagmire of logic with a general provision that appropriators use no more water than could be applied beneficially to the use for which diverted. Waste of water was "prohibited and enjoined."[13]

In addition to establishing Last Chance as one of the principal irrigation interests on the Bear River, the Dietrich decree, although emanating from a court in Idaho, also contained a provision applicable to claims in Utah and to the interstate nature of the Bear River. Judge Dietrich acknowledged that it was not his intention to establish or adjudicate the rights outside the jurisdiction of his court. His goal, impressive to this author as an extreme of fairness, was to "see that there is delivered at the Utah state line such a quantity of water as is necessary . . . to satisfy said rights in accordance with their dignity and priority as herein recognized."[14]

In addition to the policy determinations concerning the "duty of water" concept and the adjudication of rights that came directly from the decree, the litigation dramatized future needs. There developed an urgency in Idaho for adjudication of waters not covered by the Dietrich decree and for measures to harmonize the power and irrigation interests on the river. Also, the decree recorded potential problems inherent in the circumstances of an interstate stream whose waters were to be administered geographically by courts of different states.

The portion of Bear River from the Wyoming-Idaho border downstream to the Stewart Dam had been excluded from the jurisdiction of the Dietrich decree. Appropriations on this portion of the river for 618 c.f.s. of Bear River water were adjudicated on March 7, 1924, in the case, *Preston-Montpelier Irrigation Company v. Dingle Irrigation Company, et al.,* heard in the Idaho Fifth District Court.[15] The decree in this case was based in part on the legal precedents established in the *Utah Power and Light Company v. The Last Chance Canal Company, et al.* struggle of 1917–1920.

Solution to a second need traceable in time from the Bear River Water Case required amendment to the Idaho Constitution. When the constitution was ratified in 1889 (prior to Idaho statehood on July 3, 1890), the possibility of hydroelectric generation along Idaho's rivers had not yet been addressed. As originally drafted, the constitution defined beneficial uses justifying water appropriations as domestic, agricultural, and manufacturing purposes, these in the priority listed.[16] In 1928 the constitution was amended to authorize specifically state regulation of the use of Idaho water for power purposes.[17]

Certain difficulties, actual or potential, relating to administration of water rights and other matters on an interstate stream, such as the Bear River, had first arisen officially in the Dietrich decree. Problems of this nature were finally addressed thirty-five years later when, on February 4, 1955, the state of Idaho ratified an interstate compact relating to the waters of the Bear River among the states of Idaho, Utah, and Wyoming. This was the Bear River Compact.[18] This compact, however, was not in effect until 1958 owing to delays in ratification by certain of the other parties to the agreement.[19] The "major purposes" of the compact were "to remove the causes of present and future controversy over the distribution and use of the water of the Bear River; to provide for efficient use of water for multiple purposes; to permit additional development of the water resources of Bear River; and to promote interstate comity."

The Bear River Commission, the joint commission to administer the compact, consisted of ten members—three from each participating state and one nonvoting member as chairman and representing the United States. Two of the Idaho commissioners were, by law, required to be "residents within the watershed of the Bear River in Idaho."[20] Among the functions of the Bear River Commission were determination, when necessary, of the existence of a water emergency and establishment of water delivery schedules that would remain in effect during such emergency. The water delivery schedules, derived in part from public hearings, were to "recognize and incorporate . . . priority of water rights as adjudicated in each of the signatory states."

The compact also prescribed the establishment of "a reserve for irrigation" in the Bear Lake Reservoir. Waters of the lake below 5,912.91 feet were assigned a priority for irrigation and were intended as sufficient to meet any requirements for supplemental irrigation water at a scale as had been experienced during the droughts of the 1930s.

The Bear River Compact was amended after twenty years on December 22, 1978,[21] in accordance with existing review and amendment procedures. A purpose "to accomplish an equitable apportionment of the waters of the Bear River among the compacting states" was added. Also added was authorization for storage of Bear River waters in Utah and Wyoming upstream from the Stewart Dam, provided that the level of Bear Lake did not fall below 5,911 feet. The Last Chance Canal Company strongly opposed acceptance of the 5,911 feet value on the basis that such upstream storage could reduce the irrigation reserve and "would seriously lower the level of Bear Lake in a series of dry years and thus jeopardize the downstream vested water rights." The canal company first proposed a minimum level for Bear Lake of 5,916.17 feet, then took a compromise position at 5,914 feet, but to no avail. It was important to the Last Chance system that their irrigators "have assurance that . . . [they] can continue to rely upon the availability of storage water in Bear Lake at a reasonable cost."[22]

The Bear River Compact and the Bear River Commission have been of utmost importance to all interests in Bear River waters. The significance of the Last Chance Canal Company, as a historical contributor to water distribution policy and as a continuing principal in irrigation matters, has been recognized by membership on the Bear River Commission. Fred M. Cooper, secretary of the canal company, signed the original compact and was a member of the initial commission. Don W. Gilbert, canal company president, signed the 1979 amended compact and remains a commission member.

Last Chance influence on matters of water policy has continued over the years. Alertness has been maintained to any "late-comer" claimants (such as in the instance which resulted, in 1977, in the case *Hirschi, et al. v. Utah Power and Light Company and the Last Chance Canal Company*) whose activities might become detrimental in any way to adjudicated rights on the Bear River. Hand-in-glove with such measures opposing late claims is support for any proposals which may in the future improve the Last Chance position concerning availability of water for irrigation. Application of this policy frequently placed Last Chance in an unfamiliar position of support of the old antagonist, UP&L, for new dam and reservoir proposals.

Always the innovator, during the period 1980 through early 1984 Last Chance planned, built, operated briefly, and then sold a hydroelectric plant, a development that was the "first time in Idaho"[23] under relatively new federal legislation. The goal—and result—of several years of deep involvement in this hydroelectric venture was a secure position relative to an adequate supply of irrigation water.

The Last Chance Canal Company and the many individuals associated with it since its inception almost a century ago are important features in Western history. The irrigation system they built and operated made lasting contributions to the economic viability of Gem Valley and southeastern Idaho. Many changes in water policy—some minor and some of far-reaching importance—resulted from the experience and conflict on the Bear River between irrigation and hydroelectric interests. The willingness, indeed the determination, of the founders of the Last Chance Canal Company, and those who followed them, to fight these battles should assure them an important place in history for their influence on Western water policy.

Notes

See Bibliography for full citations.

INTRODUCTION

1. McBride, "Bear Lake Country," 1:28.

2. Ibid.

3. Ibid.

4. Trenholm and Carley, *The Shoshonis,* 73.

5. Hafen and Rister, *Western America,* 241–44.

6. Irving, *Adventures of Captain Bonneville,* 197-98.

7. Hafen and Rister, *Western America,* 242.

8. Goetzmann, *Army Exploration,* 92, 108.

9. Arrington, *Great Basin Kingdom,* 41.

CHAPTER ONE: THE BEAR RIVER WATERSHED

1. UP&L History, 365.

2. Houghton, *Great Basin Story,* 205.

3. John F. MacLane, "Reply Brief on Final Hearing," October 10, 1919, 89, U.S. District Court, UP&L v. Last Chance.

4. *Grace Citizen,* October 6, 1983.

5. Houghton, *Great Basin Story,* 204.

6. UP&L History, Statement A, 134.

7. Houghton, *Great Basin Story,* 205.

8. UP&L History, Statement A, 134.

9. MacLane, "Reply Brief on Final Hearing," 89; U.S. District Court, "Transcript of Proceedings," 36, 39, 173.

10. U.S. District Court, "Transcript of Proceedings," 185.

11. UP&L History, Statement A, 134.

12. MacLane, "Reply Brief on Final Hearing," 92–93.

13. Houghton, *Great Basin Story,* 204; Judge Frank S. Dietrich, "Decree," July 14, 1920, p. 7, U.S. District Court, UP&L v. Last Chance.

14. Houghton, *Great Basin Story,* 206.

15. Irving, *Adventures of Captain Bonneville,* 197.

16. Houghton, *Great Basin Story,* 205.

17. MacLane, "Reply Brief on Final Hearing," 78.

18. UP&L History, 51, 365.

19. Simmons and Varley, *"Gems,"* 38.

20. UP&L History, 49; ibid., Statement A, 36.

CHAPTER TWO: INTRODUCTION TO IRRIGATION

1. Arrington, *Great Basin Kingdom,* 41.

2. Dunbar, *Forging New Rights,* 9–11.

3. Houghton, *Great Basin Story,* 236.

4. Dunbar, *Forging New Rights,* 12.

5. Ibid., 12–15.

6. Smythe, *Conquest of Arid America,* 182.

7. Powell, *Report on the Lands of the Arid Region,* 21.

8. Smythe, *Conquest of Arid America,* 60.

9. Dunbar, *Forging New Rights,* 46–47.

10. Smythe, *Conquest of Arid America,* 263.

11. Powell, *Report on the Lands of the Arid Region,* 131–33.

12. James, *Story of the United States,* xiv.

13. Smythe, *Conquest of Arid America,* 264.

14. James, *Story of the United States,* xvi.

15. Dunbar, *Forging New Rights,* 36.

16. Lawrence B. Lee, "Introduction to the 1969 edition," in Smythe, *Conquest of Arid America,* xxxiii.

17. James, *Story of the United States,* xvi; Smythe, *Conquest of Arid America,* 42.

18. Lee, "Introduction," in Smythe, *Conquest of Arid America,* xxxi.

19. Smythe, *Conquest of Arid America,* 269–70.

20. Dodds, *Hiram Martin Chittenden,* 26.

21. Dunbar, *Forging New Rights,* 39–40.

22. Dodds, *Hiram Martin Chittenden,* 25; Hodge, *Aridity and Man,* 107.

23. Dodds, *Hiram Martin Chittenden,* 36; Smythe, *Conquest of Arid America,* 272.

24. Dodds, *Hiram Martin Chittenden,* 31; Smythe, *Conquest of Arid America,* 271.

25. Lee, "Introduction," in Smythe, *Conquest of Arid America,* xxix.

26. Smythe, *Conquest of Arid America,* 273.

27. Ibid., 294–300.

28. Dunbar, *Forging New Rights,* 52.

29. Ibid., 55.

30. Hodge, *Aridity and Man,* 108.

CHAPTER THREE: IDAHO WATER POLICY

1. Idaho Territory, "Water Rights and Irrigation," Title IX in *Revised Statutes* (1887), 373–79.

2. Idaho, *General Laws* (1895), 174–83.

3. Idaho, *General Laws* (1899), 380–87.

4. Hutchins, "Idaho Law," 8, contains a summary of this law, as does Idaho, Application for Permit, "Instructions."

5. Idaho, *Constitution,* Sec. 3.

6. Hutchins, "Idaho Law," 2.

7. Ibid., 3.

8. Ibid., 20.

9. Idaho, *General Laws* (1895), Sec. 4.

10. Idaho, *Constitution,* Sec. 3.

11. Hutchins, "Idaho Law," 7.

12. Idaho, *Constitution,* Sec. 3.

13. Idaho, *General Laws* (1899), Sec. 5.

14. Idaho, *General Laws* (1895), Sec. 4; Idaho, Application for Permit, "Instructions," no. 3.

15. Hutchins, "Idaho Law," 21.

16. Idaho, *General Laws* (1899), Sec. 7.

17. Idaho, *General Laws* (1895), Sec. 13.

18. Hutchins, "Idaho Law," 20.

19. Ibid., 26.

20. Ibid., 23.

21. Ibid., 85.

22. Idaho, *Constitution,* Sec. 3.

23. Idaho, *General Laws* (1895), Sec. 3.

24. Idaho, *General Laws* (1899), Sec. 32.

25. Idaho, *Revised Statutes* (1887), Sec. 3165.

26. Hutchins, "Idaho Law," 38.

27. Ibid., 39.

28. Ibid., 2.

29. Idaho, *Revised Statutes* (1887), Sec. 3190.

30. Idaho, Application for Permit, "Instructions," no. 7.

31. Hutchins, "Idaho Law," 40.

32. Ibid., 41.

33. Idaho, *General Laws* (1895), Sec. 23; Idaho, *General Laws* (1899), Sec. 15.

CHAPTER FOUR: EARLIEST IRRIGATION EFFORTS

1. Simmons and Varley, *"Gems,"* 11–59.

2. Ibid., 12, 38.

3. Ibid., 39.

4. John J. Trappett, "Deposition," October 1918, U.S. District Court, UP&L v. Last Chance, 13.

5. "A Bit of Idaho Pioneering," 9.

6. J.H. Peterson, Last Chance Canal Co., "Cross Complaint," n.d., U.S. District Court, UP&L v. Last Chance, 4.

7. Simmons and Varley, *"Gems,"* 73; Trappett, "Deposition," 13, 15.

8. Peterson, "Cross Complaint," 5.

9. Trappett, "Deposition," 3, 30.

10. "A Bit of Idaho Pioneering," 9; Simmons and Varley, *"Gems,"* 73; Trappett, "Deposition," 2–3.

11. Trappett, "Deposition," 16.

12. "A Bit of Idaho Pioneering," 9.

13. Trappett, "Deposition," 4, 34.

14. Peterson, "Cross Complaint," 4.

15. Simmons and Varley, *"Gems,"* 73.

CHAPTER FIVE: THE "LAST CHANCE" IRRIGATION PROJECT

1. Simmons and Varley, *"Gems,"* 73.

2. John J. Trappett, "Deposition," October 1918, U.S. District Court, UP&L v. Last Chance, 5.

3. Ibid., 7.

4. Ibid., 8.

5. Ibid., 5.

6. Ibid., 26.

7. "A Bit of Idaho Pioneering," 9; Simmons and Varley, *"Gems,"* 73; Trappett, "Deposition," 5.

8. Last Chance Irrigation Co., "Notice of Water Right," U.S. District Court, UP&L v. Last Chance; Simmons and Varley, *"Gems,"* 74.

9. Trappett, "Deposition," 6.

10. Last Chance Irrigation Co., "Notice of Water Right."

11. Ibid.

12. Idaho, *General Laws* (1895), Sec. 4.

13. Last Chance Irrigation Co., "Notice of Water Right."

14. Trappett, "Deposition," 11.

15. Ibid., 11, 19.

16. Ibid., 19.

17. Ibid., 21.

18. Ibid., 19.

19. "A Bit of Idaho Pioneering," 9.

20. Simmons and Varley, *"Gems,"* 75.

21. "A Bit of Idaho Pioneering," 9.

22. Trappett, "Deposition," 15.

23. Simmons and Varley, *"Gems,"* 38.

24. Ibid., 75.

25. Trappett, "Deposition," 7.

26. Ibid., 29.

27. Idaho, *General Laws* (1895), Sec. 5.

28. Trappett, "Deposition," 7, 29.

29. Ibid., 6.

30. Ibid., 16.

31. Simmons and Varley, *"Gems,"* 73.

32. Ibid., 75.

33. Trappett, "Deposition," 32.

34. Sorenson, "Personal Scrap-book."

35. Simmons and Varley, *"Gems,"* 75.

36. Trappett, "Deposition," 6.

37. Ibid., 19.

38. Ibid., 9.

39. Ibid., 19.

40. Ibid., 30.

41. Simmons and Varley, *"Gems,"* 73, 80.

42. Ibid., 75.

43. Trappett, "Deposition," 8.

44. Ibid., 9.

45. Last Chance Canal Co., "Notice of Water Appropriation," U.S. District Court, UP&L v. Last Chance.

46. Idaho, *General Laws* (1899), Sec. 6.

47. Trappett, "Deposition," 24–26.

48. Ibid., 27.

49. Ibid., 24.

50. Ibid., 23.

51. Ibid., 12.

52. Ibid., 19.

53. Ibid., 24.

54. Ibid., 22.

55. Ibid., 9.

56. Ibid., 10.

57. Ibid., 9.

58. Ibid., 29.

59. Ibid., 9.

60. Simmons and Varley, *"Gems,"* 75.

61. "Quit Claim Deed," October 22, 1901, U.S. District Court, UP&L v. Last Chance.

62. Trappett, "Deposition," 21.

63. Ibid., 15–16.

64. Ibid., 16.

65. Ibid., 29.

66. Simmons and Varley, *"Gems,"* 75.

67. Trappett, "Deposition," 18.

68. "A Bit of Idaho Pioneering," 9–10.

69. Trappett, "Deposition," 9.

70. John F. MacLane, "Reply Brief on Final Hearing," October 10, 1919, 16, U.S. District Court, UP&L v. Last Chance.

71. Sorenson, "Personal Scrap-book."

72. Ibid.

73. MacLane, "Reply Brief on Final Hearing," 12.

74. Ibid., 29.

75. J. H. Peterson, Last Chance Canal Co., "Cross Complaint," n.d., p. 3, U.S. District Court, UP&L v. Last Chance.

76. Sorenson, "Personal Scrap-book."

77. MacLane, "Reply Brief on Final Hearing," 24.

78. Trappett, "Deposition," 13.

79. Ibid., 18.

80. MacLane, "Reply Brief on Final Hearing," 17.

81. "A Bit of Idaho Pioneering," 6.

82. Simmons and Varley, *"Gems,"* photograph caption, 72.

83. Trappett, "Deposition," 10.

84. Ibid.

85. Last Chance Canal Co., "Minutes," August 17, 1916.

86. Ibid., September 2, 1916.

87. Ibid., November 28, 1916; February 5, 13, June 4, July 5, 1917.

88. Last Chance Canal Co., "Map, Main Canal," April 7, 1919, U.S. District Court, UP&L v. Last Chance.

89. *Grace Citizen,* 24 April 1980.

90. Last Chance Canal Co., "Minutes," February 5, August 13, 1917.

91. Ibid., August 21, 1916.

92. Ibid., June 4, 1917.

93. Bench Canal Co., "Application for Permit to Appropriate the Public Waters of the State of Idaho," no. 7249, August 9, 1909, and no. 5771, December 31, 1909, U.S. District Court, UP&L v. Last Chance.

94. Tanner Canal Co., "Application for Permit to Appropriate the Public Waters of the State of Idaho," no. 9303, July 29, 1910, U.S. District Court, UP&L v. Last Chance.

95. Bench Canal Co., "Application for Amendment to Permit to Appropriate the Public Waters of the State of Idaho," no. 5308, March 20, 1916, U.S. District Court, UP&L v. Last Chance.

96. Tanner Canal Co., "Proof of Completion of Works," September 23, 1915, U.S. District Court, UP&L v. Last Chance.

97. Bench Canal Co., "Application for Extension of Time for Beneficial Use Proof," August 21, 1918, and "Approval of State Engineer," August 26, 1918, U.S. District Court, UP&L v. Last Chance.

98. MacLane, "Reply Brief on Final Hearing," 30.

99. Ibid., 28.

100. Last Chance Canal Co., "Minutes," May 10, 1917.

101. Sorenson, "Personal Scrap-book;" Simmons and Varley, *"Gems,"* 80–85.

102. MacLane, "Reply Brief on Final Hearing," 14.

103. Ibid., 29.

104. Peterson, "Cross Complaint," 7–10.

105. MacLane, "Reply Brief on Final Hearing," 30; Judge Frank S. Dietrich, "Decree," July 14, 1920, p. 16, U.S. District Court, UP&L v. Last Chance.

106. Last Chance Canal Co., "Minutes," December 2, 1926; April 7, 1927; April 11, 1928; February 4, 1929.

107. Sorenson, "Personal Scrap-book."

108. Last Chance Canal Co., "Minutes," November 5, 1945; September 10, November 3, 1947; August 4, 1948.

109. Simmons and Varley, *"Gems,"* 79.

110. Last Chance Canal Co., "Minutes," March 29, 1948.

111. Ibid., October 5, 1947.

112. Ibid., June 23, 1977.

113. Ibid., November 8, 1943.

114. Ibid., May 21, 1924; March 23, 1929; November 7, 1938; November 7, 1946; November 2, 1958.

115. Ibid., May 21, 1924.

116. Walt Lenhart, telephone interview with author, Logan, Utah, October 12, 1983. Mr. Lenhart formerly farmed in the area once served by the Central Canal.

117. Last Chance Canal Co., "Minutes," November 3, 1969; November 2, 1970; November 1, 1971; September 22, 1972.

118. Ibid., June 30, 1919; February 2, 1920; June 20, 1924; July 22, 1931; October 30, 1966; February 28, 1977.

119. Ibid., November 7, 1949.

120. Caribou County Water Resources Committee, "Report," 7.

121. Last Chance Canal Co., "Minutes," September 22, November 3, 1977.

122. Caribou County Water Resources Committee, "Report," 23.

123. Last Chance Canal Co., "Minutes," December 30, 1920; February 10, 1925; April 19, 1980.

124. Ibid., November 4, 1923; May 21, 1924; April 30, 1929; November 5, 1956; October 22, 1979.

125. Ibid., June 10, June 29, September 22, November 6, November 20, 1972.

126. Ibid., January 31, 1978; March 5, 1979.

127. Ibid., November 3, 1975; January 4, February 9, 1977.

128. Ibid., March 24, 1977.

129. Idaho Water Resource Board, "Soda Springs Dam Feasibility Study," June 1981, 1.

130. *Grace Citizen,* September 3, 1981.

131. Fred M. Cooper to Mrs. Sessions, 6. Mr. Cooper served as secretary, Last Chance Canal Co. from 1928 until his death in 1961. Simmons and Varley, *"Gems,"* 79.

CHAPTER SIX: RESERVOIR AND POWER DEVELOPMENTS

1. UP&L History, 41; ibid., Statement A, 90.

2. Bailey, *L.L. Nunn,* 61–62.

3. Pearson, "Federal Reclamation," 1.

4. U.S. District Court, "Transcript of Proceedings," 180.

5. Pearson, "Federal Reclamation," 1.

6. Bailey, *L.L. Nunn,* 61, 270; UP&L History, Statement A, 96.

7. Bailey, *L.L. Nunn,* 72; UP&L History, 50.

8. UP&L History, 45.

9. Bailey, *L.L. Nunn,* 69.

10. Bailey, *L.L. Nunn,* 70; UP&L History, Statement A, 120.

11. L. L. Nunn, "Deposition to John F. MacLane," March 1, 1918, U.S. District Court, UP&L v. Last Chance, 20–23.

12. P. N. Nunn, "Deposition to John F. MacLane," 14, U.S. District Court, UP&L v. Last Chance.

13. U.S. District Court, "Transcript of Proceedings," 34.

14. William Story, "Deposition to John F. MacLane," 5, U.S. District Court, UP&L v. Last Chance.

15. Lucien L. Nunn et al., "Notice of Appropriation of Water," March 24 and April 12, 1902, U.S. District Court, UP&L v. Last Chance.

16. P. N. Nunn, "Deposition to John F. MacLane," 17, U.S. District Court, UP&L v. Last Chance.

17. William Story, "Deposition to John F. MacLane," 4, U.S. District Court, UP&L v. Last Chance.

18. P. N. Nunn, "Deposition to John F. MacLane," 16, U.S. District Court, UP&L v. Last Chance.

19. W. C. Cates, "Deposition to John F. MacLane," 9, 11, U.S. District Court, UP&L v. Last Chance.

20. William Story, "Deposition to John F. MacLane," 7, U.S. District Court, UP&L v. Last Chance.

21. L. L. Nunn, "Deposition to John F. MacLane," 22, U.S. District Court, UP&L v. Last Chance.

22. Idaho Water Resource Board, "Preliminary Report," 3–28.

23. Ibid., 3–28; Pearson, "Federal Reclamation," 1; UP&L History, 135.

24. William Story, "Deposition to John F. MacLane," 7, U.S. District Court, UP&L v. Last Chance.

25. Ibid.

26. Idaho Water Resource Board, "Preliminary Report," 3–28.

27. William Story, "Deposition to John F. MacLane," 7, U.S. District Court, UP&L v. Last Chance.

28. UP&L History, 366, ibid., Statement A, 138; Simmons and Varley, *"Gems,"* 89.

29. Engineer, State of Idaho, "Water License No. 1747," November 14, 1917, U.S. District Court, UP&L v. Last Chance.

30. UP&L History, Statement I, 71–87.

31. Arthur Jobson, "Deposition," March 7, 1918, U.S. District Court, UP&L v. Last Chance. Mr. Jobson was the constructing superintendent of Telluride Power Company until Fall 1912, and of Phoenix Construction Company until March 1913.

32. Idaho Water Resource Board, "Preliminary Report," 3–28.

33. Ibid., 3–3, 3–29.

34. Judge Frank S. Dietrich, "Decree," July 14, 1920, p. 7, U.S. District court, UP&L v. Last Chance.

35. Idaho Water Resource Board, "Preliminary Report," 3–3, 3–4.

36. L. L. Nunn, "Deposition to John F. MacLane," 23, U.S. District Court, UP&L v. Last Chance.

37. Pearson, "Federal Reclamation," 2.

38. UP&L History, 41.

39. Engineer, State of Idaho, "Water License No. 8962," March 14, 1918, p. 1, U.S. District Court, UP&L v. Last Chance.

40. L. L. Nunn, "Deposition to John F. MacLane," 22, U.S. District Court, UP&L v. Last Chance.

41. Idaho Water Resource Board, "Preliminary Report," 3–29.

42. Utah Power and Light Co. and Utah-Idaho Sugar Co., "Contract," December 30, 1912, 5 (hereafter cited as "Contract").

43. L. L. Nunn, "Deposition to John F. MacLane," 22, U.S. District Court, UP&L v. Last Chance.

44. Idaho Water Resource Board, "Preliminary Report," 3–28.

45. "Contract," 11.

46. Ibid., 6.

47. Ibid., 11.

48. Idaho Water Resource Board, "Preliminary Report," 4–17.

49. UP&L History, 41, 330.

50. Ibid., 337.

51. Ibid., Statement I, 71–87.

52. Simmons and Varley, *"Gems,"* 90.

53. Ibid., 94.

54. Dietrich, "Decree," 14–24.

55. UP&L History, Statement I, 71–87.

CHAPTER SEVEN: THE BEAR RIVER WATER CASE

1. John F. MacLane, "Reply Brief on Final Hearing," 4, U.S. District Court, UP&L v. Last Chance.

2. Hutchins, "Idaho Law," 85.

3. Last Chance Canal Co., "Minutes," 28 June 1917.

4. W. G. Swendsen, "Certification of true and correct copy of Notice of Water Right, Last Chance Irrigation Co., filed with State Engineer, Boise, Idaho, March 3, 1897," U.S. District Court, UP&L v. Last Chance.

5. Last Chance Canal Co., "Minutes," July 4, 1917.

6. Ibid., February 4, August 12, 1918; February 2, December 30, 1920; February 7, December 31, 1921; February 6, 1922.

7. *Idaho Statesman,* October 3, 4, 5, 1930.

8. J. H. Peterson to William Reynolds, Clerk, District Court, Boise, Idaho, July 3, 1917, U.S. District Court, UP&L v. Last Chance.

9. W. G. Swendsen, "Deposition," March 11, 1918, U.S. District Court, UP&L v. Last Chance.

10. Clerk, District Court, Boise, Idaho, to Morris and Callister, Salt Lake City, Utah, 10 October 1918, U.S. District Court, UP&L v. Last Chance.

11. Judge Frank S. Dietrich, "Order Appointing Water Commissioner," July 12, 1918, U.S. District Court, UP&L v. Last Chance.

12. Richards and Haga to F. S. Dietrich, June 6, 1919, U.S. District Court, UP&L v. Last Chance.

13. J. H. Peterson to John Williams, president, Last Chance Canal Co., June 20, 1919. Copy provided by secretary, Last Chance Canal Co.

14. J. H. Peterson, Last Chance Canal Co., "Cross Complaint," n.d., U.S. District Court, UP&L v. Last Chance, 4–5.

15. John J. Trappett, "Deposition," October 1918, p. 4, U.S. District Court, UP&L v. Last Chance.

16. Ibid., 6.

17. Swendsen, "Notice of Water Right."

18. Last Chance Canal Co., "Notice of Water Appropriation," U.S. District Court, UP&L v. Last Chance.

19. Trappett, "Deposition," 19.

20. Last Chance Canal Co., "Notice of Water Appropriation."

21. "Quit Claim Deed," October 22, 1901, U.S. District Court, UP&L v. Last Chance.

22. Peterson, "Cross Complaint," 6.

23. Sorenson, "Personal Scrap-book."

24. MacLane, "Reply Brief on Final Hearing," 16.

25. Ibid., 12.

26. Ibid., 12, 29.

27. Trappett, "Deposition," 9.

28. Ibid., 13.

29. Peterson, "Cross Complaint," 3.

30. Bench Canal Co., "Application for Permit to Appropriate the Public Waters of the State of Idaho," no. 7249, August 9, 1909, and no. 5771, December 31, 1909, U.S. District Court, UP&L v. Last Chance.

31. Tanner Canal Co., "Application for Permit to Appropriate the Public Waters of the State of Idaho," no. 9303, July 29, 1910, U.S. District Court, UP&L v. Last Chance.

32. Last Chance Canal Co., "Minutes," May 10, 1917.

33. W. G. Swendsen to W. O. Creer, Lund, Idaho, April 11, 1919. Copy provided by secretary, Last Chance Canal Co.

34. MacLane, "Reply Brief on Final Hearing," 28.

35. W. O. Creer to W. G. Swendson [*sic*], April 17, 1919. Copy provided by secretary, Last Chance Canal Co.

36. MacLane, "Reply Brief on Final Hearing," 14, 29.

37. Trappett, "Deposition," 24.

38. Peterson, "Cross Complaint," 10.

39. Ibid., 7.

40. Ibid., 11.

41. Ibid., 12.

42. Judge Frank S. Dietrich, "Decree," July 14, 1920, pp. 14-113, U.S. District court, UP&L v. Last Chance.

43. Lucien L. Nunn et al., "Notice of Appropriation of Water," March 24 and April 12, 1902, U.S. District Court, UP&L v. Last Chance.

44. W. C. Cates, "Deposition to John F. MacLane," March 1, 1918, 9, U.S. District Court, UP&L v. Last Chance.

45. Idaho Water Resource Board, "Preliminary Report," 3–28.

46. W. C. Cates, "Deposition to John F. MacLane," 9, 11, U.S. District Court, UP&L v. Last Chance.

47. Idaho Water Resource Board, "Preliminary Report," 3–28.

48. Arthur Jobson, "Deposition," March 7 1918, p. 5, U.S. District Court, UP&L v. Last Chance.

49. Idaho Water Resource Board, "Preliminary Report," 3–3, 3–4.

50. Simmons and Varley, *"Gems,"* 89.

51. Engineer, State of Idaho, "Water License No. 1747," November 14, 1917, U.S. District Court, UP&L v. Last Chance.

52. UP&L History, Statement I, 71–87.

53. Utah Power and Light Co. and Utah-Idaho Sugar Co., "Contract," December 30, 1912, 5 (hereafter cited as "Contract").

54. Idaho Water Resource Board, "Preliminary Report," 3–29.

55. Dietrich, "Decree," 23.

56. "Contract," 5–6.

57. Ibid., 3.

58. Nunn, "Notice of Appropriation of Water," March 24, 1902.

59. MacLane, "Reply Brief on Final Hearing," 6.

60. Ibid., 5.

61. Ibid., 26.

62. Ibid., 7, 10, 15.

63. Ibid., 10.

64. Ibid., 11–16, 25.

65. Ibid., 19–21.

66. Ibid., 18–19.

67. Ibid., 28–29.

68. Ibid., 29.

69. Idaho, *Revised Statutes* (1887), Sec. 3165.

70. Idaho, Application for Permit, "Instructions," no. 7.

71. MacLane, "Reply Brief on Final Hearing," 39, 72, 73.

72. Ibid., 38–77.

73. Ibid., 30–31.

74. Ibid., 39.

75. Ibid., 60.

76. Ibid., 65.

77. Ibid., 77.

78. Photographs of fields of George Telford farm, J. E. Williams farm, Edward Kirby farm, July 16, 1918; of Louisa Medford farm, July 17, 1918; and of lateral canal serving George Dalton and others, June 25, 1919, U.S. District Court, UP&L v. Last Chance.

79. MacLane, "Reply Brief on Final Hearing," 69–70.

80. Ibid., 46.

81. Ibid., 33.

82. Ibid., 82–83.

83. Ibid., 85, 87.

84. Ibid., 93–94.

85. Judge Frank S. Dietrich to Margaret Franklin, Boise, Idaho, October 17, 1925, U.S. District Court, UP&L v. Last Chance.

86. John F. MacLane to Budge and Merrill et al., Pocatello, Idaho, May 24, 1920, U.S. District Court, UP&L v. Last Chance.

87. Peterson and Coffin to John F. McLane [*sic*], Salt Lake City, Utah, June 10, 1920, U.S. District Court, UP&L v. Last Chance.

88. Dietrich, "Order Appointing Water Commissioner," June 24, 1920, U.S. District Court, UP&L v. Last Chance.

89. John F. MacLane, "Plaintiff's Bill of Costs," July 12, 1920, U.S. District Court, UP&L v. Last Chance.

90. Dietrich, "Decree," 117.

91. MacLane, "Plaintiff's Bill of Costs."

92. Dietrich, "Decree," 7.

93. Ibid., 12–13.

94. MacLane, "Reply Brief on Final Hearing," 4.

95. Dietrich, "Decree," 16–17.

96. Ibid., 113.

97. Ibid., 10.

98. Ibid., 11.

99. Ibid., 116.

100. Ibid., 117.

101. Ibid., 9.

102. Ibid., 14.

103. Ibid., 7–8.

104. Ibid., 8–9.

105. Ibid., 8.

106. Ibid., 7–8.

107. Ibid., 14–24.

108. Ibid., 19.

109. Ibid., 14–24.

110. C. E. Tappan to Judge F. S. Dietrich, Boise, Idaho, January 28, 1921, U.S. District Court, UP&L v. Last Chance.

111. Last Chance Canal Co., "Minutes," November 4, 1923.

112. Orrin Harris, secretary, Last Chance Canal Co., telephone interview with author, Grace, Idaho, March 4, 1984.

113. W. O. Creer to Last Chance Canal Co., Grace, Idaho, February 15, 1922. Copy provided by secretary, Last Chance Canal Co.

114. Last Chance Canal Co., "Minutes," October 13, 1922; February 5, 1923.

115. Ibid., February 5, 1923.

116. Sorenson, "Personal Scrap-book."

117. Simmons and Varley, *"Gems,"* 78.

118. Utah Power and Light Co. to George Telford, Grace, Idaho, November 29, 1924. Copy provided by secretary, Last Chance Canal Co.

CHAPTER EIGHT: CHALLENGES AND SOLUTIONS

1. Russell D. Stoker to author, February 10, 1984.

2. J-U-B Engineers, Inc., "Feasibility Study," 14.

3. J-U-B, "Feasibility Study," 14–15.

4. Ibid., 14.

5. F. M. Cooper, secretary, Last Chance Canal Co., to Governor C. Ben Ross, Boise, Idaho, July 27, 1934. Copy provided by secretary, Last Chance Canal Co.

6. UP&L History, Statement A, 36.

7. *Grace Citizen,* December 21, 1978.

8. Ibid., January 29, 1979.

9. Cooper to Governor Ross.

10. *Grace Citizen,* December 21, 1978; January 25, 1979.

11. Cooper to Governor Ross.

12. Idaho Water Resource Board, "Soda Springs Dam," June 1981, 1; *Grace Citizen,* October 13, 1983.

13. UP&L History, 41, 330.

14. Last Chance Canal Co., "Minutes," December 30, 1920.

15. Simmons and Varley, *"Gems,"* 78–79.

16. Last Chance Canal Co., "Minutes," February 10, 1925.

17. Simmons and Varley, *"Gems,"* 79.

18. J-U-B, "Feasibility Study, 3.

19. Ibid., 45.

20. *Grace Citizen,* July 29, 1982.

21. J-U-B, "Feasibility Study."

22. Last Chance Canal Co., "Minutes," April 19, 1980.

23. J-U-B, "Feasibility Study," 3.

24. Ibid., 13, Appendix A-2.

25. Ibid., 1, 4; Last Chance Canal Co., "Minutes," September 2, November 3, 1980.

26. J-U-B, "Feasibility Study," 23.

27. *Grace Citizen,* September 18, 1980; February 26, 1981.

28. J-U-B, "Feasibility Study, 8–11, 23.

29. Ibid., 1, 16.

30. *Grace Citizen,* September 18, 1980; February 26, 1981.

31. J-U-B, "Feasibility Study, 11, 18.

32. Ibid., 1, 45.

33. *Grace Citizen,* November 13, 1980.

34. Ibid., February 19, 1981.

35. J-U-B, "Feasibility Study, 13, Appendix A-6.

36. *Grace Citizen,* July 2, 1981; Last Chance Canal Co., "Minutes," November 2, 1981.

37. *Grace Citizen,* July 2, 1981; Last Chance Canal Co., "Minutes," November 2, 1981.

38. Last Chance Canal Co., "Minutes," May 1, June 4, November 2, 1981.

39. *Grace Citizen,* March 18, July 22, November 18, 1982.

40. Ibid., July 2, 1981.

41. Ibid., January 20, 1983.

42. Ibid., April 24, 1982.

43. Ibid., March 18; November 11, 18, 1982.

44. Ibid., July 1, 22, 1982.

45. Ibid., November 18, 1982.

46. Ibid., January 20, 1983.

47. Ibid., December 15, 1983.

48. Ibid., November 11, 1982.

49. Ibid., August 12, 1982; December 15, 1983.

50. Ibid., August 12, 1982.

51. Ibid., December 15, 1983.

52. Ibid., January 26, 1984.

53. Ibid., December 15, 1983.

54. Orrin Harris, telephone interview with author, March 5, 1984.

CHAPTER NINE: AN ASSESSMENT

1. John J. Trappett, "Deposition," October 1918, pp. 13-14, U.S. District Court, UP&L v. Last Chance.

2. Ibid., 19.

3. Fred M. Cooper to Mrs. Sessions, October 11, 1956, as contained in "A Bit of Idaho Pioneering," 6.

4. Smythe, *Conquest of Arid America,* 191–93.

5. Dunbar, *Forging New Rights,* 41.

6. Smythe, *Conquest of Arid America,* 312.

7. Cooper to Mrs. Sessions, 6.

8. Sorenson, "Personal Scrap-book."

9. UP&L History, 51, 365.

10. UP&L History, Statement A, 135.

11. Judge Frank S. Dietrich, "Decree," July 14, 1920, U.S. District Court, UP&L v. Last Chance.

12. Idaho, Application for Permit, "Instructions," no. 7.

13. Dietrich, "Decree," 113.

14. Ibid., 12–13.

15. Idaho Water Resource Board, "Preliminary Report," 4–7.

16. Idaho, *Constitution,* Sec. 3.

17. Hutchins, "Idaho Law," 1.

18. Idaho, *General Laws* (1955), chap. 218.

19. Dunbar, *Forging New Rights,* 145.

20. Idaho, *General Laws* (1955), chap. 219.

21. Idaho, *General and Special Laws* (1979), chap. 322.

22. Randall C. Budge, Last Chance Canal Co., as quoted in *Grace Citizen,* December 21, 1978.

23. *Grace Citizen,* December 15, 1983.

Bibliography

Arrington, Leonard J. *Great Basin Kingdom*. Cambridge, Mass.: Harvard University Press, 1958.

"A Bit of Idaho Pioneering." Unpublished history of John Ira Allsop. Copy of portions of this unpublished manuscript was provided to author by Elaine Johnson, Caribou County Historical Society, Soda Springs, Idaho, who had received it from Mrs. Dick Smith, granddaughter of John Ira Allsop.

Bailey, Stephen A. *L. L. Nunn: A Memoir*. Ithaca, N.Y.: Cayuga Press, 1933.

Caribou County Water Resources Committee. "Report: Caribou County Water Resources." March 1, 1968. Office of the County Agent, Caribou County, Soda Springs, Idaho.

Chittenden, Hiram Martin. *A History of the American Fur Trade of the Far West*. 1902. Reprint. Stanford, Calif.: Academic Reprints, 1954.

Dodds, Gordon B. *Hiram Martin Chittenden*. Lexington: University Press of Kentucky, 1973.

Dunbar, Robert G. *Forging New Rights in Western Waters*. Lincoln: University of Nebraska Press, 1983.

Goetzmann, William H. *Army Exploration in the American West*. New Haven, Conn.: Yale University Press, 1959.

Hafen, LeRoy R. and Carl Coke Rister, *Western America*. New York: Prentice-Hall, Inc., 1941.

Harris, Orrin. Secretary, Last Chance Canal Co., Grace, Idaho. Interviews with author on June 18, 1982, October 11, 1983, and March 4, 5, 1984.

Hodge, Carle, ed. *Aridity and Man*. Washington, D.C.: American Association for Advancement of Science, no. 74, 1963.

Houghton, Samuel G. *A Trace of Desert Waters: The Great Basin Story*. Glendale, Calif.: Arthur H. Clark Co., 1976.

Hutchins, Wells A. "Idaho Law of Water Rights." *Idaho Law Review* 5 (Fall 1968): 1-129.

Idaho. "Application for Permit to Appropriate the Public Waters of the State of Idaho." This is an administrative form to be completed by prospective appropriators after the enactment of an applicable statute in 1903. The "Instructions" for completion of the form contain a valuable summary of the 1903 Idaho water law.

Idaho. *Constitution.*

Idaho. *General Laws* (1895), (1899), (1955).

Idaho. *General and Special Laws* (1979).

Idaho Statesman (Boise).

Idaho Territory. *Revised Statutes* (1887).

Idaho. Water Resource Board, "Preliminary Report: Bear River Basin Investigation," February 1970. Copy of portions of this report was provided to author by Ralph J. Mellin, Idaho Dept. of Water Resources, Boise, Idaho.

Idaho. Water Resource Board. "Soda Springs Dam Feasibility Study." June 1981. This report was provided to author by Orrin Harris, secretary, Last Chance Canal Co., Grace, Idaho.

Irving, Washington. *The Adventures of Captain Bonneville.* 1837. Reprint. Portland, Oreg.: Binfords and Mort, Publ., 1950.

James, George Wharton. *The Story of the United States Reclamation Service.* New York: Dodd, Mead and Co., 1917.

J-U-B Engineers, Inc., Pocatello, Idaho. "Feasibility Study for the Last Chance Canal Co., Grace, Idaho, to the U.S. Dept. of Energy, Small Hydroelectric Project." March 1981. Copy of this report is available at the office of the secretary, Last Chance Canal Co., Grace, Idaho.

Last Chance Canal Co. "Correspondence File (Historical)." Office of the secretary, Last Chance Canal Co., Grace, Idaho.

————"Minutes, Meeting of the Board of Directors," August 7, 1917. Office of the secretary, Last Chance Canal Co., Grace, Idaho.

McBride, Rayola. "Bear Lake County." In Idaho Poets' and Writers' Guild, *The Idaho Story.* Iona, Idaho: Ipas Publishing Co., 1968, 1:28–30.

Pearson, Denton. "Federal Reclamation and Power Projects at Bear Lake." Memorandum to Reid Nielson, Dept. of Interior, Regional Solicitor, Salt Lake City, Utah: n.d. Copy of this memorandum provided to author by Keith R. McCarthy, U.S. Dept. of Interior, Washington, D.C.

Powell, John Wesley. *Report on the Lands of the Arid Region of the United States: U.S. Geographical and Geological Survey of the Rocky Mountain Region.* 1879. Reprint. Cambridge, Mass.: Harvard University Press, 1962.

Simmons, Vivian and Ruth Varley, eds. *"Gems" of our Valley.* Providence, Utah: Keith W. Watkins and Son, Inc., 1977.

Smythe, William E. *The Conquest of Arid America.* 1900. Reprint. Seattle: University of Washington Press, 1969. Introduction to the 1969 Edition by Lawrence B. Lee.

Sorenson, Alice T. "Personal Scrap-book." Public Library, Grace, Idaho.

Stoker, Russell D. (Bear River Watermaster) to author, February 10, 1984.

Contains fifteen enclosures with hydrological data pertaining to Bear Lake and Bear River.

Trenholm, Virginia Cole and Maurine Carley, *The Shoshonis: Sentinels of the Rockies*. Norman: University of Oklahoma Press, 1964.

UP&L History. Utah Power and Light Co. "History of Origin and Development." Salt Lake City, Utah: May 11, 1937. This document was prepared in connection with Federal Power Commission Request Order. It consists of the basic report and appendixes, Statements A to I, inclusive. Copy is available at Special Collections Library, Utah State University, Logan, Utah.

U.S. District Court, District of Idaho, Eastern Division. Equity no. 203. Utah Power and Light Co. versus Last Chance Canal Co. Adjudicated July 14, 1920. Container 37024, Box 154, Federal Records Center, Seattle, Wash. A wide variety of documents has been obtained from this archive, and the documents are cited whenever possible by author or title in chapter endnotes.

———"Transcript of Proceedings," 12–16 June 1919, Utah Power and Light Co. versus Last Chance Canal Co. et al. Copy provided to author at office in Pocatello, Idaho, of Randall C. Budge, attorney, Last Chance Canal Co.

Utah Power and Light Co. and Utah-Idaho Sugar Co. "Contract." December 30, 1912. Copy of this document provided to author by Ralph J. Mellin, Idaho Dept. of Water Resources, Boise. Copy is also available at the office of the Recorder, Box Elder County, Brigham City, Utah.